HARCOURT HORIZONS

Michigan

State Activity Book

Orlando Austin Chicago New York Toronto London San Diego

Visit *The Learning Site!*
www.harcourtschool.com

REVIEWERS

Carol Bacak-Egbo
Social Studies Consultant
Waterford Schools
Waterford, MI

Michael R. Federspiel
Coordinator of Social Studies
Midland Public Schools
Midland, MI

Dr. M. Jean Ferrill
Professor of Geography
Department of Geography
Northern Michigan University
Marquette, MI

Dr. Charles K. Hyde
Professor of History
Department of History
Wayne State University
Detroit, MI

Dr. Gary Manson
Professor of Geography
Department of Geography
Michigan State University
East Lansing, MI

Heather Reid
Social Studies
Department Head
Loy Norrix High School
Kalamazoo, MI

Dr. Renay M. Scott
Associate Professor
Teacher Education and
 Professional Development
Central Michigan University
Mount Pleasant, MI

Karen R. Todorov
Social Studies Education
 Consultant
Michigan Department of
 Education
Lansing, MI

Copyright © by Harcourt, Inc.

All rights reserved. No part of this publication may be reproduced or transmitted in any form or by any means, electronic or mechanical, including photocopy, recording, or any information storage and retrieval system, without permission in writing from the publisher.

Permission is hereby granted to individual teachers using the corresponding student's textbook or kit as the major vehicle for regular classroom instruction to photocopy copying masters from this publication in classroom quantities for instructional use and not for resale. Requests for information on other matters regarding duplication of this work should be addressed to School Permissions and Copyrights, Harcourt, Inc., 6277 Sea Harbor Drive, Orlando, Florida 32887-6777. Fax: 407-345-2418.

HARCOURT and the Harcourt Logo are trademarks of Harcourt, Inc., registered in the United States of America and/or other jurisdictions.

Printed in the United States of America

ISBN 0-15-335735-5

1 2 3 4 5 6 7 8 9 10 082 10 09 08 07 06 05 04 03 02

Contents

Introduction
- 6 What Is History?
- 7 What Does a Historian Do?
- 8 The Tools of a Historian
- 9 Time and Time Lines
- 10 Primary and Secondary Sources
- 11 The Six Themes of History
- 13 Core Democratic Values
- 14 What Do You Know About Michigan?

·UNIT· 1

An Introduction to Michigan
- 16 Lesson 1: Where in the World Is Michigan?
- 18 Activity 1: Mapping Your Community
- 19 Lesson 2: Regions in Michigan
- 21 Activity 2: Comparing Regions Over Time
- 22 Lesson 3: Rivers and Lakes in Michigan
- 24 Activity 3: Michigan's Ports
- 25 Lesson 4: The Climate of Michigan
- 27 Activity 4A: Comparing Temperature and Precipitation in Michigan
- 29 Activity 4B: The Lake Effect
- 30 **Unit 1 Practice Test**

·UNIT· 2

Michigan's Earliest People
- 32 Lesson 5: The First Americans
- 33 Activity 5: The Old Copper Indians
- 34 Lesson 6: The Hopewell Culture
- 35 Activity 6: Farm or Save the Past?
- 36 Lesson 7: Michigan's Native American Groups
- 38 Activity 7: Working in an Ojibwa Village
- 40 Lesson 8: Anishabek and the People of the Three Fires
- 41 Activity 8: Build a Confederation
- 42 **Unit 2 Practice Test**

UNIT 3

Colonization and Settlement

- 44 Lesson 9: The French Arrive
- 46 Activity 9A: The Founding of Detroit
- 47 Activity 9B: Fort Pontchartrain
- 48 Lesson 10: The Fur Trade
- 49 Activity 10: Be a Fur Trader
- 50 Lesson 11: The French and Indian War
- 51 Activity 11: Examine Native American Alliances
- 52 Lesson 12: Pontiac's Rebellion
- 53 Activity 12: The Proclamation of 1763
- 54 **Unit 3 Practice Test**

UNIT 4

Americans Come to Michigan

- 56 Lesson 13: Michigan and the American Revolution
- 58 Activity 13: Examine the Declaration of Independence
- 60 Lesson 14: The Constitution of the United States
- 61 Activity 14: Write a Constitution
- 62 Lesson 15: Forming the Northwest Territory
- 64 Activity 15: Townships in the Northwest Territory
- 65 Lesson 16: Michigan Becomes a Territory
- 66 Activity 16: Michigan's Changing Borders
- 68 Lesson 17: Pioneer Life in Michigan
- 69 Activity 17A: Changing Views on Settlement
- 70 Activity 17B: Getting to the Frontier
- 71 Lesson 18: The War of 1812
- 72 Activity 18: Causes and Effects of the War of 1812
- 73 Lesson 19: Statehood for Michigan
- 74 Activity 19A: The Toledo War
- 75 Activity 19B: Michigan State Government
- 76 **Unit 4 Practice Test**

UNIT 5

The New State of Michigan

- 78 Lesson 20: Growth as a State
- 80 Activity 20A: Mining in Michigan
- 81 Activity 20B: Lumbering in Michigan
- 82 Lesson 21: Slavery and Freedom
- 83 Activity 21: Sojourner Truth
- 84 Lesson 22: Michigan in the Civil War
- 86 Activity 22A: The Battle of Gettysburg
- 87 Activity 22B: Women in the Civil War
- 88 Lesson 23: Cities and Industries Grow in Michigan
- 90 Activity 23A: Building an Automobile Industry in Michigan
- 92 Activity 23B: Effects of the Automobile
- 93 Lesson 24: The Cereal Bowl of America
- 94 Activity 24: From Corn to Corn Flakes
- 96 **Unit 5 Practice Test**

UNIT 6

Challenges for Michigan

- 98 Lesson 25: A New Century Brings Changes
- 99 Activity 25: Detroit and the Great Migration
- 100 Lesson 26: The Great Depression
- 101 Activity 26: Living Through the Great Depression
- 103 Lesson 27: The Rise of Unions
- 104 Activity 27: Work for Change
- 105 Lesson 28: World Events Affect Michigan
- 106 Activity 28A: The Arsenal of Democracy
- 107 Activity 28B: Rosie the Riveter
- 108 Lesson 29: The Mackinac Bridge
- 109 Activity 29: Michigan's Highways
- 110 Lesson 30: The Civil Rights Movement in Michigan
- 111 Activity 30: The Freedom March
- 112 **Unit 6 Practice Test**

UNIT 7

Modern Michigan

- 114 Lesson 31: Michigan at the End of the Twentieth Century
- 116 Activity 31A: A City Grows Again
- 117 Activity 31B: The Michigan Beverage Container Act
- 118 Lesson 32: Who Lives in Michigan?
- 120 Activity 32A: Newcomers to Michigan
- 121 Activity 32B: Community Heritage
- 122 Lesson 33: Where Do People Live in Michigan?
- 124 Activity 33: Michigan's Population
- 125 Lesson 34: Michigan's Economy Today
- 127 Activity 34A: The Service Industry in Michigan
- 128 Activity 34B: The Automobile Industry in Michigan Today
- 129 Lesson 35: Michigan Looks to the Future
- 130 Activity 35A: Predict Michigan's Future
- 131 Activity 35B: Discover the History of Your Community
- 132 **Unit 7 Practice Test**

Thinking Organizers

- 134 Questions About History
- 135 People, Events, and Ideas
- 136 Main Idea and Supporting Details
- 137 Compare and Contrast
- 138 Causes and Effects
- 139 Sequence
- 140 Categorize
- 141 Time Line
- 142 Outline Map of Michigan
- 143 Outline Map of the United States

What Is History?

Studying **history**, or what happened in the past, helps people understand how events of the past affect the present. By studying history, we can find links between past and present events. These links can help us predict how events in the present will affect the future.

Understanding history requires knowing when events took place. The order in which events take place is called **chronology** (kruh•NAH•luh•jee). **Historians**, or people who study the past, look closely at the chronology of events. This helps them better understand how one event affects another and how the past and the present connect.

RESPOND
1. Why is it important to study history?
2. How does studying chronology help historians analyze events?

Detroit, 1794

Detroit, today

What Does a Historian Do?

How do historians learn about the past? To do it, they do many jobs. They ask questions, find evidence, identify points of view, and draw conclusions. You can do the same jobs as you study Michigan's history.

Asking Questions

Historians ask themselves questions about every event they study. Then they try to find answers to all their questions through research. This process is called **inquiry.**

Finding Evidence

Historians find **evidence**, or proof, of when, where, why, and how things happened. To do this, they rely on many different kinds of references. Historians read books and newspapers from long ago. They study old diaries, letters, and postcards. They look at paintings and photographs from the past. They also listen to oral histories. An **oral history** is a story of an event told aloud. These different kinds of evidence help historians piece together the history of people, events, and places.

(continued)

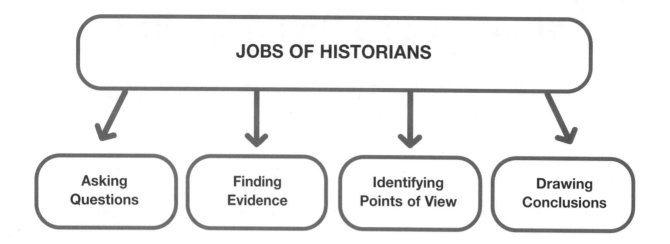

Identifying Points of View

Historians think about why different people of the past said or wrote what they did. They try to understand people's points of view. A **point of view** is how a person sees things. It can be affected by whether someone is young or old, male or female, or rich or poor. Background and experiences also affect point of view. People with different points of view may have different ideas about the same event.

Drawing Conclusions

After historians have identified the facts about a historical event, they still have work to do. They need to analyze the event. To **analyze** an event is to examine each part of it and relate the parts to each other. Analyzing an event lets historians draw conclusions about how and why it happened. They add up the historical facts about a topic and think about what they already know about the topic. Then they can draw conclusions about the topic.

RESPOND

1. What is meant by the term *inquiry*?
2. What are four examples of historical evidence?
3. Why might two people have different points of view about the same event?
4. How might historians use different pieces of evidence to draw a conclusion about an event?

Time and Time Lines

To understand the history of Michigan or any other place, you need to know when important events happened. A time line can help you with this. A **time line** is a diagram that shows the order in which events took place and the length of time between them. Putting events in the order in which they took place can help you understand how one event may have led to another.

The time line below shows when some important events in the early history of Michigan took place. The earliest date in a time line is at the left end of the time line. The most recent date is at the right end. Like a map, a time line has a scale. But the marks on a time line's scale show units of time, not distance.

Time lines can show different units of time. Some show events that took place during one day, one month, or one year. Others show events that took place during a **decade**, or a period of ten years. On the time line below, the space between the dates on the top stands for one **century**, or 100 years. A time line can also show a **millennium**, a period of 1,000 years. The year 2001 marked the beginning of a new millennium.

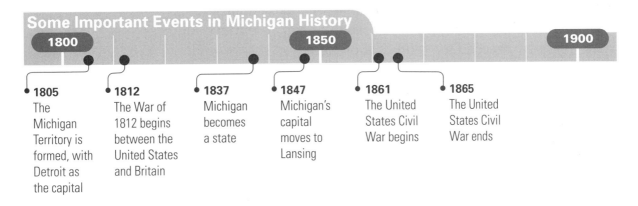

RESPOND

1. How is a millennium different from a century?
2. In what year did the Michigan Territory form?
3. Which happened first, the War of 1812 or the Civil War?
4. How many years passed between Detroit becoming the capital of the Michigan Territory and Lansing becoming the state capital?

Primary and Secondary Sources

People who study the past look for evidence, or proof. They want to be sure they know what really happened long ago. They look for evidence in two different kinds of sources—primary sources and secondary sources.

Primary sources are records made by people who saw or took part in an event. They may have written their thoughts in journals or diaries. They may have told their stories in letters, poems, or songs. They may have given speeches. They may have painted pictures or taken photographs. All these primary sources are records made by people who saw what happened. Objects made or used during an event can also be primary sources.

Secondary sources are records made by people who were not at an event. Books written by authors who only heard about or read about an event are secondary sources. So are magazine articles and newspaper stories written by people who did not take part in the event. Paintings or drawings by artists who did not see the event are also secondary sources.

RESPOND

1. How is a primary source different from a secondary source?
2. Why do historians use both primary sources and secondary sources?
3. Why might two secondary sources show different points of view?
4. Why might historians compare several people's letters and diaries to help them draw conclusions about an event?

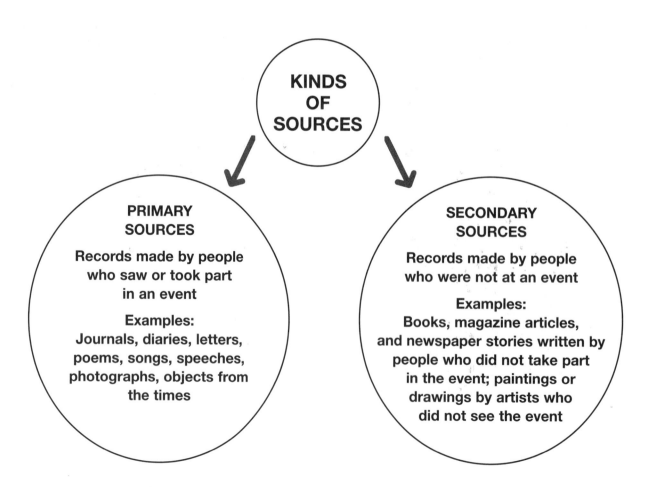

The Six Themes of History

The six themes of history can help you make connections between the past and the present. They can also help you predict what might happen in the future. Think about the six themes as you study Michigan's history.

Theme 1
Civilization, Cultural Diffusion, and Innovation

THEME: INTERACTIONS

The movement of people, the spread and interaction of cultures, and new ideas, or innovations, in technology continue to shape Michigan. As you read about different groups that have settled in Michigan, think about how their languages, religions, and other cultural ideas have affected the state. Also pay attention to how inventions such as the automobile continue to influence Michigan.

Theme 2
Human Interaction with the Environment

THEME: GEOGRAPHY

The geography and natural resources of Michigan greatly influence the development of the state. As you read, think about geographical features, such as landforms and bodies of water, that make Michigan unique. Also think about how people change the environment and how the environment changes people.

Theme 3
Values, Beliefs, Economics, Political Ideas, and Institutions

THEME: VALUES AND NEW IDEAS

Values and new ideas have influenced change and continue to affect the people of Michigan. While reading the lessons and activities in this book, look for examples of how individuals and groups share values and new ideas.

(continued)

Theme 4
Conflict and Cooperation

THEME: CONFLICT AND COOPERATION

Michigan has been and continues to be shaped by conflicts that can be resolved through cooperation and compromise. As you read, think about what has caused conflicts in Michigan in the past and how these conflicts have been solved.

Theme 5
Comparative History of Major Developments

THEME: COMPARISONS

The historical significance of Michigan's growth can be understood by comparing events in the state to regional, national, and world developments. While learning about Michigan's history, pay attention to related events in other parts of the region, the United States, and the world. For instance, you will read about world wars that affected Michigan's citizens and its economy.

Theme 6
Patterns of Social and Political Interaction

THEME: HUMAN EXPERIENCES

Learning about the experiences of common people creates an understanding of social change connected to immigration, migration, and the industrialization of Michigan. As you study Michigan's history, remember that your family has influenced the state's past and continues to influence its present and future. You may wish to find out why your family or ancestors moved to the state and compare their reasons with other groups' reasons.

RESPOND

1. Why do people use the six themes of history to study Michigan's past?
2. Which theme of history best connects to ways people use Michigan's environment?
3. Which theme of history best relates to conflicts over borders?
4. Which theme of history best relates to the movement of people into Michigan and to social change?
5. Which theme best covers how world events affect Michigan?

Core Democratic Values

People who study the past view events through their values and beliefs, or points of view. Americans who study history consider how events relate to the ten **core democratic values**. Understanding the core, or central, democratic values can also help people make better decisions. These democratic values are based on two documents written when the United States was formed: the Declaration of Independence and the Bill of Rights, which is part of the Constitution of the United States.

Life
This means your right to life and your responsibility to respect life.

Diversity
This means that people should respect one another's cultural and ethnic backgrounds, ways of life, and beliefs.

The Pursuit of Happiness
This means the right to seek happiness in your own way as long as you do not interfere with other people's rights.

Popular Sovereignty
This means that the power of the government comes from the people.

Liberty
This means freedom to live your life in dignity and security. It also means your ability to make personal, political, and economic choices.

Equality
This means equal opportunity and equal protection under the law for everyone.

The Common Good
This means the duty to promote the welfare of the community and to work together for the benefit of all.

Patriotism
This means that people show devotion and loyalty to their country and its values and principles.

Justice
This means that everyone should be treated fairly and that no group should have more rights than any another group.

Truth
This means that the government and the people should tell the truth and seek the truth.

RESPOND
1. What are the two documents on which the core democratic values are based?
2. How do the core democratic values affect how Americans study history?
3. When people display the American flag, which core democratic value are they honoring?
4. Which core democratic values affect your life daily? Explain.

What Do You Know About Michigan?

In this book you will learn all about Michigan, but you probably already know some facts about the state. Think about what you already know about Michigan. Then look at the map on page 15. Use what you already know and what you see on the map to answer the questions below.

RESPOND

1. Which large bodies of water border Michigan?

   _____

   _____

2. Which states border Michigan?

   _____

   _____

3. What physical feature shapes much of Michigan's borders?

   _____

4. What part of Michigan borders Lake Michigan?

   _____

5. Is Michigan north or south of the equator?

   _____

6. What is the state capital of Michigan?

   _____

7. Which river flows through the state capital?

   _____

8. How is Michigan different from most other states?

   _____

   _____

(continued)

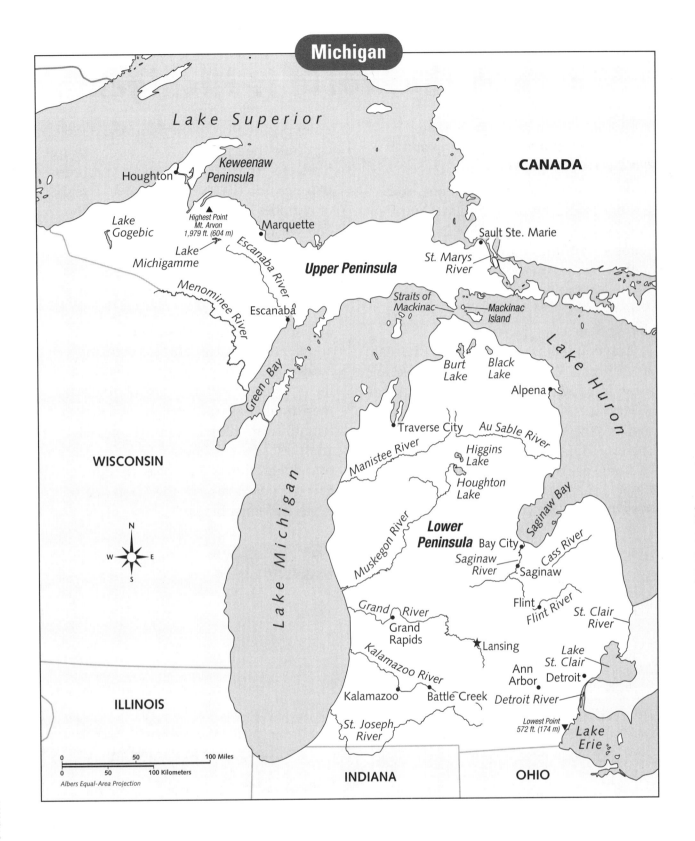

LESSON 1　　　　　　　　　　　　　　　　　　**THEME: GEOGRAPHY**

Where in the World Is Michigan?

MAIN IDEA: There are many ways to describe the location of Michigan and every place in Michigan.

How would you tell someone from another country where you live? You could say that you live in the United States, one of the countries on the continent of North America. You could say, too, that you live in the Middle West, one of the four large regions of the United States.

You could also say you live in the state of Michigan. What if someone wanted to know in which of the two large land areas of Michigan you live? To this you can answer "the Upper Peninsula" or "the Lower Peninsula."

There are other ways to describe where you live. One way is to give your absolute location. **Absolute location** is an exact position on Earth's surface. People can state their absolute location by using lines of latitude (LA•tuh•tood) and lines of longitude (LAHN•juh•tood) on a map or a globe. **Lines of latitude** run east and west and do not meet. The equator itself is a line of latitude that divides Earth into the Northern and Southern Hemispheres. Lines of latitude are measured in degrees north and south of the equator. **Lines of longitude** run north and south and meet at the North and South Poles. They are farthest apart at the equator. These lines are also called **meridians** (muh•RID•ee•uhnz). Lines of longitude are measured in degrees east and west of the prime meridian. The prime meridian is the line of longitude that divides Earth into the Western and Eastern Hemispheres.

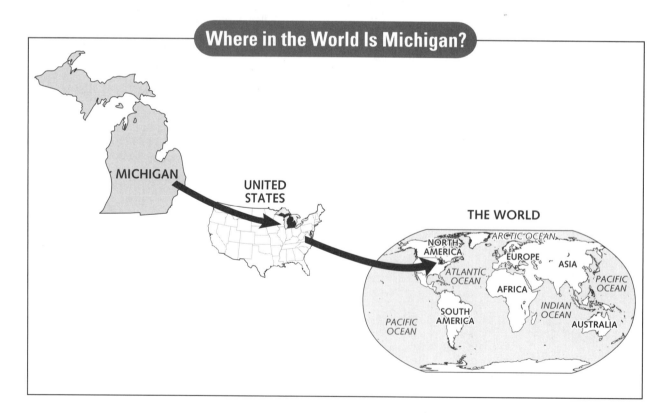

Where in the World Is Michigan?

(continued)

16　MICHIGAN STATE ACTIVITY BOOK

Together, lines of latitude and lines of longitude form a grid. You can use this grid to locate any place on Earth. You do this by naming the line of latitude and the line of longitude closest to that place. For example, the absolute location of Detroit, Michigan, is 42°N, 83°W.

You can also describe your location by telling where you live in relation to other places. When you do this, you are giving your **relative location**. Many of Michigan's people live south of Canada and north of Ohio. This is one way to describe relative location. Like many people in Michigan, you might say that you live near one of the Great Lakes. If you live in the Upper Peninsula, you live south of Lake Superior and north or west of Lake Michigan. If you live in the Lower Peninsula, you live east of Lake Michigan and west of Lake Huron and Lake Erie. Or you might say that you live near one of Michigan's rivers, such as the Kalamazoo River in the Lower Peninsula. Do you live near one of Michigan's large cities? If you live in or near Detroit, you could say that you live in or near one of the biggest cities in the United States.

All of these points of reference help describe your relative location. However, no matter where you live in Michigan, you can say that you live in a state whose motto is "If you seek a pleasant peninsula, look about you."

RECALL
1. How is absolute location different from relative location?
2. What is the absolute location of Detroit, Michigan?
3. Where is the Upper Peninsula located in relation to Lake Michigan and Lake Superior?

CRITICAL THINKING—DISCUSS OR WRITE
4. **APPLY** Look at a globe. How would you describe to someone living in Spain where Michigan is located in relation to his or her country?
5. **EVALUATION** Which do you think is more important in your daily life, knowing your absolute location or knowing how to describe your location relative to other places? Why? When might you need to know each kind of location?
6. **INQUIRY** What questions would you ask a friend in another town in Michigan to find out where that friend lives in relation to your community?

ACTIVITY 1   SKILL: MAP AND GLOBE

Mapping Your Community

DIRECTIONS: Use the map on this page to complete the activities below.

1. With an *X*, mark the location of your community on the map.

2. Write the line of latitude and line of longitude closest to where you live.

   _____

3. Describe the location of your community in relation to three physical features or cities in the state.

   _____

   _____

   _____

   _____

4. Fill in the lines on the "postcard" to explain where you live. If necessary, review the information on pages 16 and 17 of this book.

WHERE I LIVE IN THE WORLD

Name _____

Street Address _____

City or Town _____

State _____

Country _____

Continent _____

---

18 MICHIGAN STATE ACTIVITY BOOK   UNIT 1

LESSON 2

THEME: GEOGRAPHY

Regions in Michigan

MAIN IDEA: Michigan can be divided into different kinds of regions based on features such as landforms, the way people earn a living, or where people live.

To study Michigan, people often divide the state into smaller regions. A **region** is an area with at least one feature that makes it different from other areas. Regions can be based on natural features. These natural features may include landforms such as mountains, plains, hills, and valleys. Michigan is often divided into two land regions: the Superior Upland and the Great Lakes Plains.

The Superior Upland lies in the western part of Michigan's Upper Peninsula, bordering Lake Superior. Forests cover many of the hills in this region of Michigan. The mineral-rich Porcupine Mountains rise in the northwestern part of this region. The highest point in the state, Mount Arvon, is in the Superior Upland. This peak rises 1,979 feet (603 m) above sea level.

The land drops in **elevation**, or height, from the western half of Michigan's Upper Peninsula to the eastern half. This eastern half of the Upper Peninsula is part of the region called the Great Lakes Plains. Swamps and limestone hills make up the land along Lake Michigan. Sandstone cliffs lie along Lake Superior.

The Great Lakes Plains also cover all of the Lower Peninsula of Michigan. Unlike most of the land in the Superior Upland, most of the land in the Great Lakes Plains in the Lower Peninsula is flat. There are some rolling hills, however,

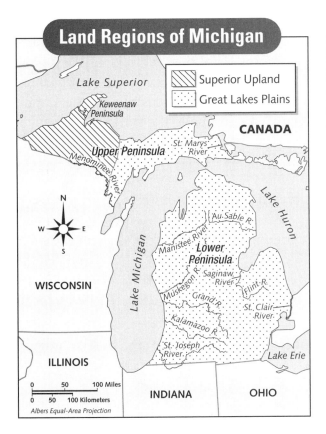

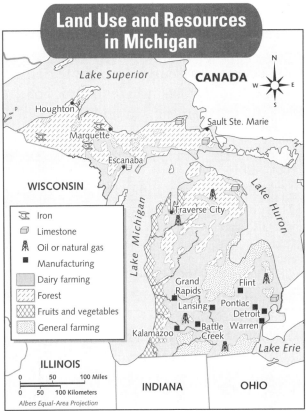

(continued)

UNIT 1

MICHIGAN STATE ACTIVITY BOOK 19

and sand dunes line the shores of Lake Michigan. One of the flattest parts of the state is in the southeast part of the Great Lakes Plains, along Lake Erie.

Other kinds of regions in Michigan are based on the **economy**, or the way people use resources to meet their needs. In the state's timber regions—the Upper Peninsula and the northeastern part of the Lower Peninsula—many people earn their living by cutting down trees for wood. In the mineral-rich Upper Peninsula, some people work at mining iron ore, limestone, and copper. If you are looking for farms in Michigan, you are most likely to find them in the central and southern parts of the Lower Peninsula. The rich soil there is good for farming. Many of Michigan's cities are in the state's manufacturing (man•yuh•FAK•chuh•ring) regions. **Manufacturing** is the making of products. Some of the products made in Michigan's factories are automobiles, cereal, and furniture.

People also divide places into regions based on where people live. In Michigan most people live in **urban**, or city, regions. These urban regions include the cities of Detroit, Flint, Lansing, Grand Rapids,

The area around Traverse City is a major cherry-producing region.

Kalamazoo, Saginaw, Bay City, Ann Arbor, and Battle Creek and their surrounding areas. Other people in Michigan live in **rural**, or country, regions where there are no large cities.

Another kind of region is one that shares a government. This kind of region is a political region. Cities and counties in Michigan are examples of the state's political regions. Political regions have exact boundaries set by law. All the residents within those boundaries live by the same local laws and have the same leaders.

RECALL
1. What are some of the different kinds of regions in which you live?
2. What are Michigan's two land regions?
3. Where is the Great Lakes Plains region in relation to the Upper Peninsula and the Lower Peninsula?

CRITICAL THINKING—DISCUSS OR WRITE
4. **ANALYZE** Why do you think more people in Michigan live in urban regions than in rural regions?
5. **INQUIRY** With a partner, write questions about how people in your area and in other parts of Michigan use resources to meet their needs. Research the answers to your questions, and present the information to the class in an oral report.
6. **REFLECT** In which land region do you live? How do the physical features around you affect your way of life?

ACTIVITY 2   SKILL: MAP AND GLOBE

Comparing Regions Over Time

DIRECTIONS: All regions develop and change over time, including regions based on where people live. The maps below show population densities in five regions in Michigan in 1910 and 2000. Population density tells how many people live in an area of a certain size. Compare the maps to see how Michigan's urban regions have changed over time. Then answer the questions that follow.

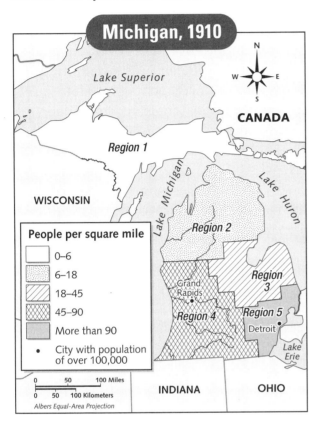
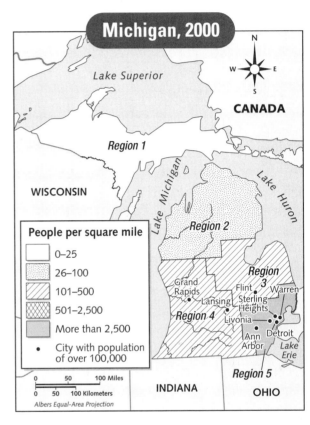

1. Which region was most densely populated in 1910? What was its population density?

2. Which region was most densely populated in 2000? What was its population density?

3. Which region of Michigan has remained the least densely populated between 1910 and 2000? What might be the reason for this?

4. How many cities with more than 100,000 people were there in Michigan in 1910? In 2000? Where are most of them located?

UNIT 1   MICHIGAN STATE ACTIVITY BOOK   21

LESSON 3 — THEME: GEOGRAPHY

Rivers and Lakes in Michigan

MAIN IDEA: Michigan's rivers and lakes are one of the state's greatest resources.

Michigan has valuable **natural resources**, or things in nature that people can use. One of the state's most important natural resources is water. Michigan's rivers and lakes provide fresh water for homes and businesses and are important to the growth of the state.

Water is everywhere in Michigan. The state borders four of the five Great Lakes. Lake Superior, Lake Michigan, Lake Huron, and Lake Erie all touch Michigan. The state's shores stretch 3,288 miles (5,291 km), forming the second-longest coastline in the United States. In addition to the four Great Lakes, Michigan has more than 11,000 inland lakes. Most of the larger lakes are in the Lower Peninsula. The largest of these is Houghton Lake.

People need fresh water to drink and use for other purposes, and the Great Lakes are Michigan's most important source of fresh water. In fact, the Great Lakes hold about one-fifth of the world's fresh water.

Water has played an important role in Michigan's growth and economic success. Today, Michigan's lakes and rivers are used as waterways for moving people and goods from place to place. The same was true in the past. Native Americans were the first to use the rivers as roads. Traveling in canoes, they carried goods to trade with other Native Americans along Michigan's rivers and along the shores of the Great Lakes. Then early French fur trappers, loggers, and settlers used Michigan's rivers and lakes to cross the land. Later, businesses used rivers to move goods such as timber through the state and to the rest of the country.

Some settlements that started on Michigan's waterways, such as Detroit, Port Huron, and Saginaw, in time became important centers of trade and manufacturing. Their locations alongside waterways allowed people to easily transport raw materials and manufactured goods to and from the settlements. These cities are now among Michigan's largest.

Today three rivers in Michigan connect the state to other parts of the world. The St. Marys, the St. Clair, and the Detroit Rivers make it possible for ships to pass from one part of the Great Lakes to another and to ports around the world.

(continued)

People were not always able to use some of Michigan's waterways as transportation routes. Not all rivers were **navigable**, or deep and wide enough for ships to use. For example, rapids on the St. Marys River blocked boats traveling from Lake Superior to the other Great Lakes. A **rapid** is a rocky place in a river where a sudden drop in elevation causes water to be fast-moving and dangerous. Beginning in 1797, engineers built canals to bypass the rapids. A **canal** is a human-made waterway. Today the Soo Canals remain important to trade on the Great Lakes. They allow ships to pass between Lake Superior and Lake Huron.

To make further use of rivers in Michigan, people have built dams across some of them. The dams hold the water back. Each one creates a human-made lake, or **reservoir** (REH•zuh•vwar). Lake Michigamme (mih•shuh•GA•mee), in the Upper Peninsula, is one of Michigan's largest reservoirs.

Water from reservoirs can be used to turn machines that create hydroelectric power. **Hydroelectric power** is electricity made by waterpower. More than 100 dams in Michigan power hydroelectric plants. Many homes and businesses depend on this hydroelectric power.

Michigan's waterways are also used for recreation, as places for people to have fun. Sailing, rafting, canoeing, swimming, and fishing are among the popular activities that people often enjoy on Michigan's lakes and rivers.

RECALL
1. In what ways do people in Michigan use rivers and lakes?
2. What rivers in Michigan make it possible for ships to travel from one of the Great Lakes to another?

CRITICAL THINKING—DISCUSS OR WRITE
3. CORE DEMOCRATIC VALUES: THE COMMON GOOD Many of Michigan's rivers flow through other states. Why might people planning to build a new dam on a river in Michigan have to think about people in other states who live along the same river?
4. EVALUATION Write a few sentences to support the following statement: The Great Lakes have helped make Michigan and the other states of the Middle West one of the largest manufacturing areas of the United States.
5. REFLECT How do you depend on Michigan's waterways in your daily life?

ACTIVITY 3

SKILL: MAIN IDEA AND SUPPORTING DETAILS

Michigan's Ports

DIRECTIONS: The main idea in a passage is the most important idea. Supporting details in a passage give more information to support the main idea. Read the passage below. Then fill in the table to give the supporting details of the main idea of each paragraph.

Michigan's rivers, especially those that flow into the Great Lakes, serve as kinds of highways. Long ago, ports were built at the mouths of some of these rivers and became important stops on the river highways. Millions of dollars' worth of goods are shipped in and out of such Michigan ports as Detroit, Sault Sainte Marie, Marquette, Saginaw, and Port Huron. So even though these ports are hundreds of miles from the Atlantic Ocean, ships leaving Michigan ports can travel all over the world.

The port of Detroit, on the Detroit River, is one of the largest ports in the United States. From the Detroit River, factories in Detroit can easily ship their goods to Lake Erie. From there the goods can be shipped to other parts of the United States and to the rest of the world. Detroit's location on the Detroit River is one reason the city is a center of the automobile industry today. The river made it easy for businesses to ship steel and automobiles in and out of the Detroit area.

Sault Sainte Marie, on the St. Marys River, is another important stop on Michigan's water highways. This city is home to the Soo Locks. These locks, which are part of the Soo Canals, are some of the busiest locks in the world. A **lock** is a section of a canal in which ships can be raised or lowered by letting water in or out. About 10,000 boats and ships pass through the locks each year to and from Lake Superior.

MAIN IDEA	→	DETAILS
1. Port cities grew along the Great Lakes.	→	A. _____ B. _____
2. Detroit is a major port city.	→	A. _____ B. _____
3. The Soo Locks are important to trade in Michigan.	→	A. _____ B. _____

LESSON 4

THEME: GEOGRAPHY

The Climate of Michigan

MAIN IDEA: The climate of Michigan affects almost every part of life in the state, including the kinds of clothing people wear and the ways people earn their living.

Compare Michigan to a state such as Arizona or Alaska. What differences in climate might you expect? Unlike those places, Michigan has a mostly moist, temperate climate. A **temperate climate** is a climate that is not too hot or too cold. In places with temperate climates, some winter days can be bitterly cold. Some summer days can be steamy and hot. However, the overall climate in the state allows people there to grow food and live comfortably for most of the year.

All of Michigan lies within the same part of North America. All places in the state are far north of the equator. Temperature and precipitation differences within Michigan are not as great as they would be between a southern state, such as Florida, and a far northern state, such as Alaska. **Precipitation** is water in the form of rain, sleet, or snow that falls to Earth's surface.

In Michigan, the greatest differences in weather occur between places in the Lower Peninsula and places in the Upper Peninsula. The **weather** in a place on a particular day is made up of the temperature, precipitation, and wind. On a January day, the temperature in Sault Sainte Marie, in the Upper Peninsula, might be around 14°F to 16°F (-10°C to -9°C). Farther south, in the Lower Peninsula, the temperature on the same day might be 26°F (3°C). The temperature during an Independence Day fireworks show in Sault Sainte Marie might be 64°F

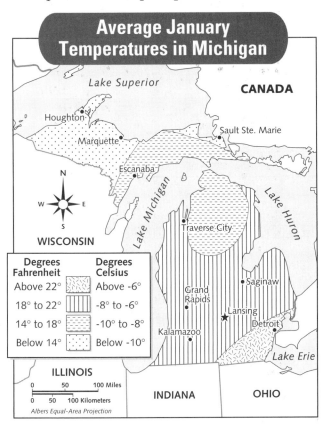

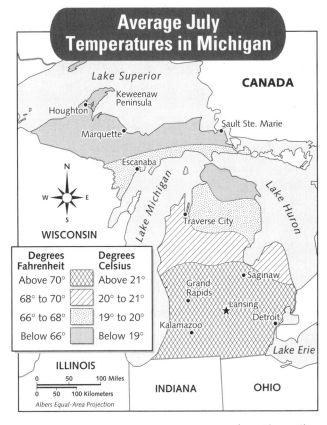

(continued)

UNIT 1
MICHIGAN STATE ACTIVITY BOOK 25

(18°C), while a thermometer in Detroit reads 73°F (22.8°C).

The differences in temperature across the state affect the way people in Michigan farm. Farmers in the Lower Peninsula have a six-month growing season. A **growing season** is the time during which the weather is warm enough for plants to grow. In contrast, the growing season in the Upper Peninsula lasts only about three months.

The Great Lakes that surround much of Michigan have a major effect on the state's climate. The westerly winds that blow across the Great Lakes cause something called the **lake effect**. It makes Michigan's weather milder than the weather in other parts of the Middle West.

Because of its northern location, Michigan receives plenty of snow in the winter. Each year, an average of about 70 inches (178 cm) of snow falls in Michigan. However, because of the lake effect, the parts of Michigan closest to the Great Lakes sometimes receive huge amounts of snow. The greatest snowfalls usually occur in the Upper Peninsula. In the Upper Peninsula's Houghton County, snowfall averages about 176 inches (447 cm) each year.

People in the Upper Peninsula have learned to adapt to all that snow. In the city of Houghton, many houses are built high off the ground so that the doors are above snow level. Homes are often built with steep roofs, so that snow slides off and falls to the ground.

RECALL
1. How do the Great Lakes affect the climate of Michigan?
2. When does the temperature in Traverse City reach about 68°F to 70°F (20°C to 21°C)?

CRITICAL THINKING—DISCUSS OR WRITE
3. **CORE DEMOCRATIC VALUES: LIFE** One of the primary responsibilities of the government is to protect the lives of the people and to ensure their safety. How do you think the Michigan state government makes sure people in Michigan are safe from dangerous weather?
4. **REFLECT** How does the climate where you live affect the way you live?
5. **INQUIRY** What do you want to learn about your local weather? Write four questions about your local weather. Keep a weather log for seven days. At the end of the period, see whether you are able to answer your questions.

ACTIVITY **4A**

SKILL: CHART AND GRAPH

Comparing Temperature and Precipitation in Michigan

DIRECTIONS: Line graphs and bar graphs can show changes over time. They can also show quantities, or amounts. Study the two graphs for Escanaba below. Then answer the questions that follow.

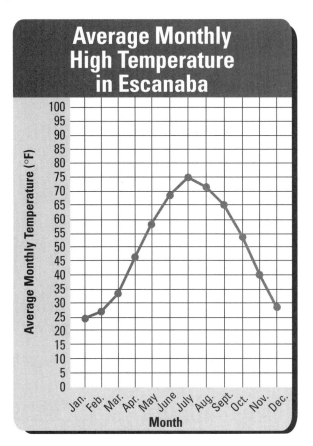
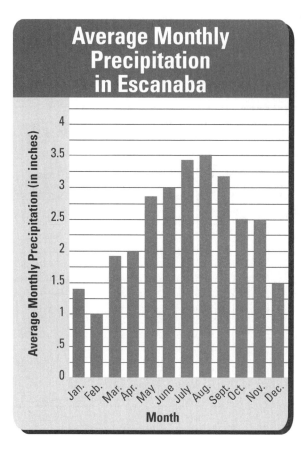

1. Which two months have the highest amount of precipitation? _____

2. What are the two warmest months in Escanaba? _____

3. What is the difference in the average January temperature and the average June temperature in Escanaba? _____

4. What is the average amount of precipitation during February in Escanaba?

_____

(continued)

UNIT 1   MICHIGAN STATE ACTIVITY BOOK   27

DIRECTIONS: The table below shows average monthly temperatures and precipitation in Detroit. Use the information in the table to complete the two graphs that follow. Use the graphs on page 27 as models.

Average Monthly Temperature and Precipitation in Detroit												
	Jan.	Feb.	Mar.	Apr.	May	June	July	Aug.	Sept.	Oct.	Nov.	Dec.
Average Monthly Temperature	23°F	26°F	36°F	47°F	58°F	68°F	72°F	70°F	63°F	51°F	40°F	28°F
Average Monthly Precipitation	1.9 in.	1.9 in.	2.5 in.	3 in.	3 in.	3.5 in.	3.1 in.	3.1 in.	3.3 in.	2.2 in.	2.7 in.	2.5 in.

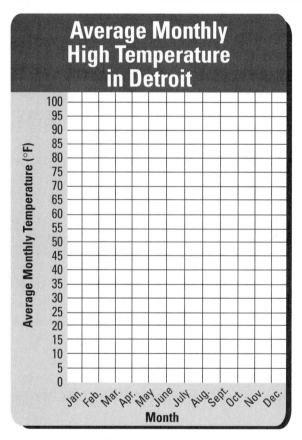

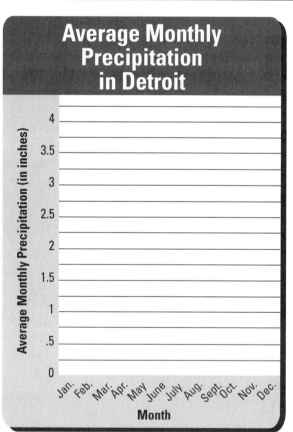

DIRECTIONS: Compare the average high temperatures and the average amount of precipitation in Escanaba and Detroit. Escanaba is in the southern Upper Peninsula. Detroit is in the southeastern Lower Peninsula. The Great Lakes are close to both cities and help shape their climates. On the lines below, compare the climates of these two Michigan cities.

_____

_____

_____

ACTIVITY **4B**

SKILL: CHART AND GRAPH

The Lake Effect

DIRECTIONS: Data such as snowfall amounts can be shown in a bar graph. Read the passage below, and study the graph. Then answer the questions that follow.

In the winter, the lake effect causes huge amounts of snow to fall in sections of the states bordering the Great Lakes, including Michigan, Minnesota, Illinois, and New York. At that time, the water in the Great Lakes is usually warmer than the air. As winds blowing from the west pass over the huge lakes, they pick up moisture from the warm water. This causes heavy snow clouds to form.

The lake effect is greatest in the northwestern Upper Peninsula and least in the southeastern Lower Peninsula. In the western Upper Peninsula, about 160 inches (406 cm) of snow fall each year, compared with 40 inches (102 cm) in the southeastern Lower Peninsula. In the winter of 1978–1979, a record amount of snow fell in Houghton County, which is in the Keweenaw Peninsula, near Lake Superior. Snowplows had to clear 350 inches (889 cm) of snow from the roads!

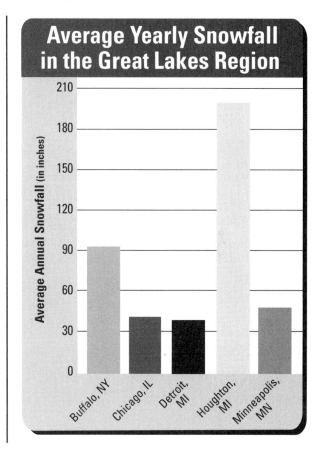

1. On average, about how much more snow falls in the city of Houghton than in Detroit? _____

2. In the winter of 2001–2002, the snowfall recorded in the Keweenaw Peninsula was 184 inches (467 cm). What is the difference between that amount and the amount that fell in the winter of 1978–1979? _____

3. Why does more snow fall in the Keweenaw Peninsula than in Detroit?

   _____

4. How do you think the city budget for Houghton or Buffalo would be different from the budgets of cities such as Phoenix, Arizona, or Miami, Florida?

   _____

UNIT 1   MICHIGAN STATE ACTIVITY BOOK   29

UNIT 1

Practice Test

PART 1 SELECTED RESPONSE

DIRECTIONS: Study the map below and use it together with what you already know to answer the questions that follow.

1. Where is Grand Rapids in relation to Kalamazoo?
   A. north
   B. south
   C. east
   D. west

2. Most major cities in Michigan
   A. are located on the Upper Peninsula.
   B. are located on the Lower Peninsula.
   C. are located on the Canadian border.
   D. are located on the shores of Lake Huron.

3. Which city is MOST likely to have a lot of snow because of the lake effect?
   A. Kalamazoo
   B. Lansing
   C. Ann Arbor
   D. Marquette

4. Which of the following cities is MOST likely to have the warmest temperature on a July day?
   A. Detroit
   B. Sault Sainte Marie
   C. Houghton
   D. Marquette

(continued)

PART 2 EXTENDED RESPONSE PREPARATION

State Your Opinion

DIRECTIONS: Read the paragraph below, and form an opinion about the topic being discussed.

> ### The Cost of Government Services
>
> Because of the increasing costs of snow removal, a town on the Upper Peninsula cannot afford to pay for many of the services it has provided in the past. The town council must decide which programs will no longer be paid for by the town. Some council members have suggested that winter recreation programs for young people, such as skiing lessons and the ice hockey leagues, could be cut. Do you think this is a good solution? Explain why or why not.

1. What is your opinion?

   _____
   _____
   _____
   _____

2. Explain why you hold that opinion.

   _____
   _____
   _____
   _____
   _____
   _____
   _____
   _____
   _____

LESSON 5

THEME: INTERACTIONS

The First Americans

MAIN IDEA: Over time, groups of early people spread out across North and South America, becoming the first Americans.

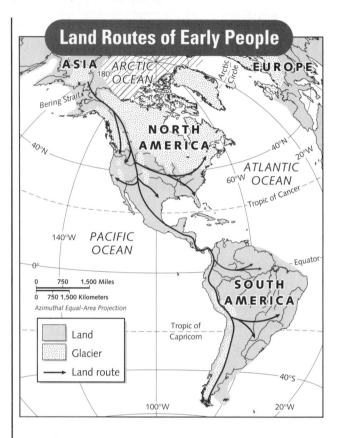

People have been living in Michigan and the rest of North America for thousands of years. Many scientists believe that the first Americans came from Asia. Today a narrow body of water called the Bering Strait separates North America from Asia. Thousands of years ago, however, huge sheets of ice called **glaciers** covered much of Earth. So much water turned into ice that the water level of the oceans dropped. At several different times this caused a "bridge" of dry land to appear where the Bering Strait is today.

Sometime between 12,000 and 40,000 years ago, people may have begun **migrating**, or moving, across the land bridge toward North America. These early people, **Paleo-Indians** (PAY•lee•oh IN•dee•uhnz), may have followed animals they were hunting. These early people were the **ancestors** of, or those who came before, today's Native Americans.

Once in North America, some people continued moving south and east. Some Paleo-Indians probably reached what is now Michigan 11,000 years ago. They were likely drawn there by the great number of animals and fish that lived near and in the Great Lakes.

Not all scientists today agree with this theory, or idea. Some think that native people may have traveled to North America in boats. However, since the first Americans did not leave behind a written history, scientists do not know exactly how the first people arrived in North America.

RECALL
1. Who were the earliest people in North America?
2. When did people probably first reach what is now Michigan?

CRITICAL THINKING—DISCUSS OR WRITE
3. **INQUIRY** What questions do you have about the first people in North America?
4. **SYNTHESIZE** Different groups of people in the Americas had different ways of life. Why do you think this happened?
5. **CORE DEMOCRATIC VALUES: DIVERSITY** Many Native Americans believe that their people have always lived in the Americas. In what ways does the right of Native Americans to hold this different point of view relate to the democratic value of diversity?

ACTIVITY 5

SKILL: IDENTIFY FACT AND OPINION

The Old Copper Indians

DIRECTIONS: A fact is a statement that can be checked and proved to be true. An opinion is a statement that describes what a person thinks or believes. Read the following passage. Then read the statements that follow. Write *Fact* next to the statements that are facts, and write *Opinion* next to the statements that are opinions.

When people first arrived in what is now Michigan 11,000 years ago, the climate of most of North America was very cold. Over the next 4,000 years, the climate slowly got warmer. During this time, animals such as the saber-toothed tiger and the mastodon became **extinct**, or no longer existed. The people who lived in the area had to change the way they lived. They started hunting many of the fish, birds, and mammals we know today. They also gathered plants and berries. Scientists call the early Americans of this time **Archaic** (ar•KAY•ik) **Indians**.

Later, copper changed the lives of Native Americans in what is now Michigan. About 5,000 years ago Native Americans learned to remove copper from the ground by heating rocks, throwing water on the rocks to crack them, and then hitting the rocks with stone hammers to free the copper metal. They hammered the copper to make metal tools, weapons, and ornaments. The Archaic Indians who used the copper are called the Old Copper Indians. Scientists believe they were among the first people to use copper.

In addition to discovering copper, the Old Copper Indians built canoes. The canoes let them get to their copper pits more easily. The canoes also allowed the Old Copper Indians to trade with Native American groups that lived far away. Scientists have found copper items from Michigan in such states as New York, Kentucky, and Illinois.

Copper fish ornament

1. _____ The Archaic Indians are more interesting than the Paleo-Indians.

2. _____ The Old Copper Indians made tools out of copper.

3. _____ Native Americans in what is now Michigan began using copper about 5,000 years ago.

4. _____ The Old Copper Indians were among the first people to use copper.

5. _____ The use of copper was the most important change in the way of life of early Native Americans in what is now Michigan.

6. _____ Copper is the most difficult metal to heat and hammer for spear points.

7. _____ The Old Copper Indians made canoes.

8. _____ The Old Copper Indians had better trade items than other early people.

Lesson 6

THEME: INTERACTIONS

The Hopewell Culture

MAIN IDEA: The ancient Hopewell people farmed, built mounds, and traded all over much of North America.

Between 2,200 and 1,300 years ago, a new culture, or way of life, rose up in the middle western and eastern parts of what is now the United States, including southwestern Michigan. Scientists today call it the Hopewell culture.

Unlike earlier Native Americans, who often moved around, the Hopewell people built villages. They were the first people in Michigan to grow corn, squash, beans, and other crops. They also fished, gathered plants, and hunted for food. To build the houses they lived in all year, they made use of the resources around them. First, they used wooden poles made from the trees in Michigan's forests to build frames. Then they covered the frames with tree bark or animal skins.

The Hopewells are often known as Mound Builders. This is because they built huge mounds of earth in which they buried their dead. Some mounds were as tall as four-story buildings. One of the largest mounds that the Hopewells built was located near where Grand Rapids is today. The Hopewells often placed pottery, carved ornaments, and other items with their dead.

Based on artifacts they have found in North America, scientists believe that the Hopewells traveled long distances to trade with other Native Americans. Among the items they traded were copper tools and weapons. In return, they brought home such goods as alligator teeth from Florida, shells from the Atlantic coast, and silver from Canada.

The Hopewells are considered to have had one of the most advanced cultures of their time. All their many achievements required the Hopewells to cooperate, or work together. For example, they had to cooperate for farming, building earthworks, and trading.

Carved stone ornaments such as this have been found in Hopewell mounds.

RECALL
1. How was the Hopewell culture different from that of earlier Native American groups in Michigan?
2. What was the purpose of the Hopewells' mounds?

CRITICAL THINKING—DISCUSS OR WRITE
3. **EVALUATION** How do you think trading with other Native Americans helped the Hopewells spread their culture? How do you think people who spoke different languages traded together?
4. **INQUIRY** How might you compare the Hopewell culture with an early Native American culture in another part of the United States, such as the Southwest? What are some questions you would want to have answered so that you could make comparisons?
5. **ANALYZE** What are three ways in which life in Michigan today is different from life at the time of the Hopewells?

ACTIVITY 6   SKILL: MAKE A THOUGHTFUL DECISION

Farm or Save the Past?

The people of the Hopewell culture sometimes built walls of earth in the form of circles, squares, and other shapes near their mounds. The low walls, or earthworks, often covered many acres. In the 1800s Michigan settlers needed farmland, so they plowed over some of the earthworks and mounds. More than 600 Native American mounds still exist in Michigan today. You can visit one of these mounds in Bronson Park, in Kalamazoo.

DIRECTIONS: In a group, evaluate the decision of some Michigan farmers in the 1800s to plow over Hopewell earthworks. Use the chart below to help you organize your thoughts. Compare your chart with those of other groups. Discuss what people today might do differently from people in the past.

PLOW OVER OR SAVE THE EARTHWORKS?	
1. What arguments might people in the 1800s have used for and against plowing over the earthworks and mounds?	**For:**
	Against:
2. What could have been other solutions to the problem?	**A.**
	B.
3. What would have been the likely effect of each solution?	**A.**
	B.
4. Do you think the decision of farmers in the 1800s to plow over the earthworks and mounds was a good one? Give your reasons.	**Explain:**

UNIT 2   MICHIGAN STATE ACTIVITY BOOK 35

LESSON 7

THEME: INTERACTIONS

Michigan's Native American Groups

MAIN IDEA: Most Great Lakes tribes lived in villages where they farmed and gathered food. They also hunted.

Beginning 1,000 years ago, several Native American tribes left their homes along the Atlantic coast of North America. By the mid-1600s, nine Native American tribes lived in the upper Great Lakes region. The Hurons (HYUR•ahnz), or Wyandots, were the largest group. Other tribes included the Ottawas (AHT•uh•wahz), the Menominees (muh•NAHM•uh•neez), the Potawatomi (paht•uh•WAHT•uh•mee), the Sauks (SAWKS), the Foxes, the Winnebagos (win•uh•BAY•gohz), the Miamis (my•AM•eez), and the Ojibwas (oh•JIB•wahz). Another name for the Ojibwas is Chippewas.

Most of the tribes lived in villages near lakes and rivers. They used these bodies of water for drinking, for fishing, and as transportation routes. The people of the region traveled in birchbark canoes on these waterways. From their canoes, many of them fished and hunted animals. Sometimes they even gathered wild grasses from the riverbanks. The most popular of the grasses was wild rice.

The people of this region made good use of the plants and animals in the area. Besides gathering wild plants, they grew crops such as corn, squash, beans, and tobacco. They ate the meat of the animals they killed. Then they turned the skins and fur of the animals into clothing, and shelters. For the most part, the women tended the crops or gathered food while the men hunted.

The Native Americans in what is now Michigan also used natural resources for their houses. Most people in the region

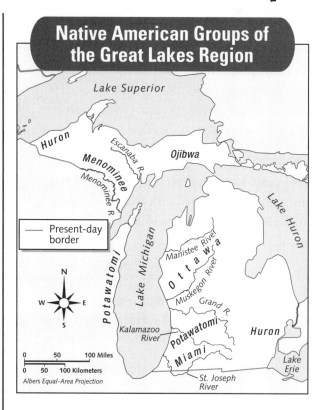

lived in wigwams. A **wigwam** is a shelter in the shape of a dome or cone. Some wigwams in the Great Lakes region were made by bending young tree limbs to form a dome. Bark covered this frame. Other wigwams were made with thin, long branches that were tied together and covered. Other people made larger rectangular, domed houses from wood poles, bark, and grasses.

Since the Great Lakes people depended on plants and animals, they showed respect for all living things. They believed that a Great Spirit lived in all parts of nature.

Some tribes, such as the Objibwas and Potawatomi, had a secret society called the Midewiwin (mih•deh•WIH•win), or Grand Medicine Society. Only someone who appeared to have a special connection with the Great Spirit could become a member. Society members would sing to the new

(continued)

member, "We are now to receive you into the Midewiwin..." The new member would sing back something like, "I have the medicine in my heart, and I am strong as a bear."

Although the tribes of the Great Lakes region were similar in many ways, each had its own way of life. The Hurons settled between Lakes Erie, Ontario, and Huron. They lived in large, walled villages for most of the year.

The Ojibwas, the second-largest group in the region, were mostly nomads. A few Ojibwas were farmers. However, most did not farm because the soil where they lived was too rocky. Instead, they traveled around Lake Superior and into Canada, gathering food.

The Ottawas fished, farmed, and hunted along the shores of Lake Huron or on nearby islands. Some of the Ottawas spent much of their time trading. They used rivers to reach trading partners.

The Menominees of the Upper Peninsula harvested more wild rice than any other tribe and usually did not farm. By trading their extra wild rice, they were able to get almost anything they wanted. The Menominees painted their clothes with interesting designs and attached porcupine quills.

The Potawatomi lived in the Lower Peninsula as well as in southern Wisconsin. In the mid-1600s a French priest described their farms: "Their country is excellently [well] adapted to raising Indian corn, and they have fields covered with it."

Some Great Lakes tribes did not live in what is today Michigan. Even so, they sometimes hunted or traded on Michigan land. The Foxes and the Sauks lived in what is now central Wisconsin. In the winter they left their villages to hunt buffalo. The Miamis, of what is now southern Wisconsin, also hunted buffalo after the harvest. The Winnebagos lived in Wisconsin. They traded buffalo robes with the Menominees for wild rice. They lived in bark lodges instead of wigwams and used dugout canoes.

There were two main language groups in the Great Lakes region—Algonquian (al•GAHN•kwee•uhn) and Iroquoian (ir•uh•KWOY•uhn). The Ojibwas, Ottawas, Potawatomi, and Menominees spoke languages in the Algonquian group. The Hurons spoke languages in the Iroquoian group.

RECALL
1. How did the Great Lakes tribes use natural resources to meet their needs?
2. In the tribes of the Great Lakes region, how was work divided among the men and women?

CRITICAL THINKING—DISCUSS OR WRITE
3. **Synthesize** Why do you think people who had land that could be farmed might decide to settle instead of traveling all the time in search of food?
4. **Inquiry** What are three questions that you might have asked an early Native American to learn more about his or her culture?
5. **Core Democratic Values: Justice** The different peoples of the Great Lakes region had different ways of life. Do you think all of the groups had a way of settling arguments? Explain.

ACTIVITY 7

SKILL: COMPARE AND CONTRAST

Working in an Ojibwa Village

Most Ojibwas did not settle in one place all year long, and most did not farm. Even so, they did live in villages during the warmer months of the year. The people living in an Ojibwa village all belonged to the same clan. A **clan** is a group of families that are all related to one another. Each clan was named after an animal. The bear clan might live in one village and the wolf clan in another.

Usually each Ojibwa clan built its village near a river. That way, the villages would have water to drink and places to fish. They would also be able to use the river for transportation.

Each family had its own wigwam in the village. The Ojibwas made their wigwams by forming frames out of tree branches and covering them with bark.

The people in a village had to work together to survive. Cooperation was needed to gather plants for food. Cooperation was also necessary for hunting and for protection from wild animals and other dangers.

In an Ojibwa village different people handled different jobs. Dividing work among different workers is called **division of labor**. Usually men and women had different kinds of jobs.

Men spent most of their time hunting and fishing to find food for the entire clan. Often they did these activities from birchbark canoes that they had made. The men used bows and arrows to kill the animals they hunted.

Women gathered food, made clothing, prepared animal hides, and cooked meals. For the most part, women were able to gather wild rice from canoes. They did this by knocking rice kernels, or seeds, from plants growing on the riverbanks into the bottom of the canoes.

If the Ojibwa clan did not live too far north or on land that was too rocky, women planted corn, beans, and squash in gardens. Since the growing season was short, crops were not a large part of the Ojibwa diet.

In the winter the Ojibwas left their villages. Individual families traveled in different directions in search of food. The men hunted moose, caribou (KAR•uh•boo), bear, and small animals. Women looked for wild berries, nuts, and roots.

Both men and women wore similar kinds of clothes. They wore buckskin tops, leggings, and moccasins. In the winter the Ojibwas also wore coats of rabbit skins or furs. The women decorated the clothing with beads and porcupine quills to make beautiful designs.

DIRECTIONS: Place the letter of each caption in the correct place on the picture. Then write a paragraph in which you compare the jobs of Ojibwa men and women.

A. Ojibwa women gather wild rice.

B. An Ojibwa man hunts wild animals.

C. A woman sews designs on leggings.

D. Women care for crops.

E. A man builds a birchbark canoe.

(continued)

UNIT 2　　　　　　　　　　　　　　　　　　　　MICHIGAN STATE ACTIVITY BOOK　39

LESSON 8　　　　　　　　　　　　　　　　　　　　THEME: INTERACTIONS

Anishabek and the People of the Three Fires

MAIN IDEA: The Ojibwas, Ottawas, and Potawatomi cooperated in a group called Anishabek, or the Three Fires Confederacy.

Most tribes in the Great Lakes region cooperated with each other in many ways including trading. Three tribes in particular had a special relationship. The Ojibwas, the Ottawas, and the Potawatomi formed a confederation. A **confederation,** or a confederacy, is a group in which the members agree to work together for their common good. The confederation in Michigan was called Anishabek, or the Three Fires Confederacy. Today many people call the group the People of the Three Fires.

The member tribes of the Three Fires Confederacy were similar in many ways. They treated each other like members of a family. The oldest members were the Ojibwas, or the "older brothers." The Ottawas were in the middle, and the Potawatomi were called the "younger brothers."

The People of the Three Fires did not have one ruler. Each tribe kept its own identity and culture, but the three tribes worked together to make decisions about problems that they all faced. In this way, they had a form of government, although there were no official laws.

Each tribe had its own strengths. Some of the Ojibwas were skilled at fishing. The Ottawas were very good traders and canoe makers. The Potawatomi were considered the first farmers of Michigan.

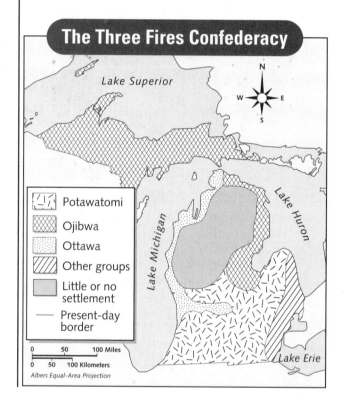

RECALL
1. What tribes make up the People of the Three Fires?
2. Who were thought of as the "older brothers" of the People of the Three Fires?

CRITICAL THINKING—DISCUSS OR WRITE
3. **APPLY** What kinds of problems do you think the People of the Three Fires discussed?
4. **CORE DEMOCRATIC VALUES: THE COMMON GOOD** How is the way in which the three tribes worked as a confederacy an example of working for the common good?
5. **APPLY** How might your school be different if it were part of a confederacy with two other schools?

ACTIVITY 8   SKILL: Solve a Problem

Build a Confederation

DIRECTIONS: In groups of three, role-play the People of the Three Fires. Make sure that each tribe is represented. Discuss how the member tribes could work together to address certain important issues. Then write an agreement for the confederacy.

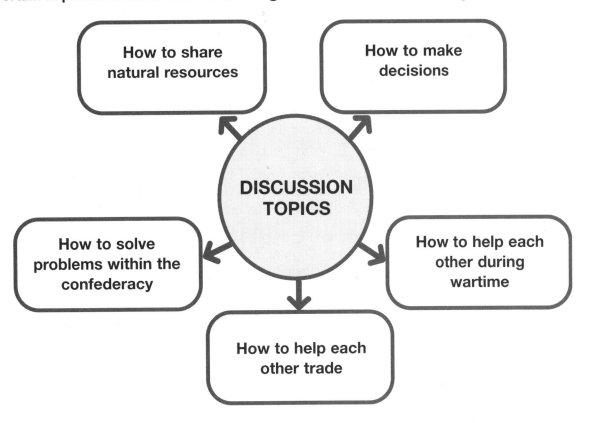

FINAL AGREEMENT:

_____

_____

_____

_____

_____

_____

_____

UNIT 2

Practice Test

PART 1 SELECTED RESPONSE

DIRECTIONS: Study the table, and use it and what you have learned in Unit 2 to answer the questions that follow.

Tribe	Language	Location	Lifestyle	Main Foods
Ojibwa	Algonquian	Near Lakes Superior and Huron and St. Marys River	Semi-settled hunters and fishers	Fish, animals, wild rice
Huron	Iroquoian	Between Lakes Erie, Ontario, and Huron	Settled farmers	Corn, beans, squash, sunflowers
Ottawa	Algonquian	Along shores of Lake Huron and on the lake's islands	Semi-settled farmers, hunters, and traders	Corn, fish, animals
Menominee	Algonquian	In Michigan's Upper Peninsula	Settled gatherers	Wild rice

1. Which of the tribes listed in the table depended MOST on a single source of food?
   A. Huron
   B. Ottawa
   C. Menominee
   D. Ojibwa

2. Why might the Ottawas have been able to survive during periods when their crops failed and the hunting and fishing were poor?
   A. They always had extra food stored away.
   B. They could trade for the food they needed.
   C. They were the best hunters of any tribe.
   D. They had friends who would give them food.

3. At a meeting of tribal leaders, which groups would probably have been able to communicate best?
   A. Ojibwa, Huron, and Ottawa
   B. Ojibwa and Menominee
   C. Huron and Ottawa
   D. Menominee, Ojibwa, and Huron

4. Which tribe's way of life was LEAST like that of the other tribes?
   A. Ottawa
   B. Menominee
   C. Ojibwa
   D. Huron

(continued)

PART 2 EXTENDED RESPONSE PREPARATION

State Your Opinion

DIRECTIONS: Read the paragraph below, and answer the questions that follow.

The Hopewell Culture

The Hopewell culture was very successful in central North America between about 2,200 years ago and 1,300 years ago. The Hopewells farmed, built large mounds of earth, and made objects from clay, metal, and stone. Their main crops were corn, squash, and beans, but they also hunted, fished, and gathered wild plants for food. The Hopewells traded with other Native American cultures all over North America. Material from the Hopewell culture has been found at sites as far apart as the Rocky Mountains, the Gulf of Mexico, and the Atlantic coast. Religion was very important to the Hopewells. Priests most likely were their rulers. The mounds they built were often burial sites. After about A.D. 500, the Hopewell culture began to decline.

1. Do you think other Native Americans may have adopted some of the ideas of the Hopewells?

   _____

2. Explain why you hold the opinion that you do.

Lesson 9

THEME: INTERACTIONS

The French Arrive

MAIN IDEA: During the 1600s, the French explored what is now Michigan, made maps of the region, set up forts and missions, and interacted with native tribes.

In the early 1600s Britain, Spain, Holland, and France started colonies in North America. A **colony** is a settlement started by people who leave their own country to live in another land. By 1608 the French had claimed most of the land that is now Canada and had started the colony known as New France.

Soon, French fur traders in the New France settlements of Quebec and Montreal wanted to expand their trade routes. The French also wanted to find a waterway to the Pacific Ocean as a shorter route to Asia. In addition, French priests wanted to teach Christianity to native peoples. For these reasons, French explorers began traveling to the Great Lakes region. In the Upper and Lower Peninsulas of Michigan, they eventually built trading posts, forts, and missions.

In 1620 Étiene Brulé (ay•tee•YEN bruh•LAY) was the first European to reach what is now Michigan. After sailing west from Quebec, he met members of the Huron tribe and traveled to the falls on the St. Marys River. By 1618 he most likely had reached the western tip of Lake Superior. In 1634 Jean Nicolet (ZHON nih•kuh•LAY) canoed from the Straits of Mackinac to Green Bay on Lake Michigan. Nicolet was searching for the Northwest Passage to the Pacific Ocean. Although he did not find a water route, he claimed the land around Lake Michigan for France.

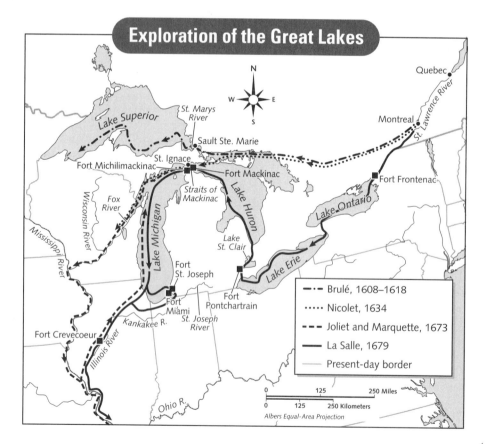

(continued)

44 MICHIGAN STATE ACTIVITY BOOK

One of the first **missions**, or religious settlements, in what is now Michigan was set up in 1660 at Keweenaw Bay. It was started by a French priest named René Ménard (reh•NAY MAY•nard). The first permanent French settlement in Michigan, however, was started in 1668 at Sault Ste. Marie. There, another priest, Jacques Marquette (ZHAK mar•KET), built a mission near some Ojibwa villages. Marquette learned native languages, hoping to teach Christianity to the native people more easily.

In 1673 Louis Joliet (zhohl•YAY), along with Marquette, led a group southward from Michigan. The men followed the shores of Lake Michigan to Green Bay. They traveled down Wisconsin rivers into the Mississippi. There they continued south for hundreds of miles before turning back.

René-Robert Cavelier, Sieur de la Salle (ren•NAY roh•bair ka•vuhl•YAY, SER deh•luh•SAL), set up Fort Miami in 1679 at the mouth of the St. Joseph River. Then, during the next three years, La Salle and his men traveled south from the Great Lakes to the mouth of the Mississippi River. La Salle claimed lands along the entire Mississippi River for France, even though Native Americans were already living there. La Salle named this region Louisiana to honor the king of France, Louis XIV.

The French continued to build forts and missions around the Great Lakes. Fort St. Joseph was set up in 1691 near the present-day city of Niles. In about 1715 the French built Fort Michilimackinac (mih•shuh•lee•MA•kuh•naw) at the Straits of Mackinac. Huron and Ottawa villages, a mission, and homes for settlers were all near the fort.

In the summers, Native American fur trappers and French fur traders met at Fort Michilimackinac to exchange their goods. As a result of trading, some of the tribes of the Great Lakes region came to own and use European goods, such as guns, metal pots and tools, mirrors, and beads.

RECALL
1. Who were two of the French explorers who helped to expand New France?
2. Where was the first permanent French settlement in what is now Michigan?

CRITICAL THINKING—DISCUSS OR WRITE
3. **APPLY** Why do you think the French wanted to expand their fur trade in North America and find a water route to the Pacific Ocean?
4. **EVALUATION** How do you think the culture of the Hurons and other Native American groups changed as a result of trading with the French? Do you think the change was good or bad? Explain your reason.
5. **REFLECT** If you were an explorer in the 1600s, what would you have liked or disliked about exploring the Great Lakes region?

ACTIVITY **9A** SKILL: ANALYZE PRIMARY SOURCES

The Founding of Detroit

DIRECTIONS: Read the passage and study the historical map. Then answer the questions.

In 1701 Antoine de la Mothe Cadillac (an•TWAHN deh luh MOHT KA•duhl•ak) got permission from the French king to build a new fort in the upper Great Lakes region. The French hoped the fort would protect their land claims in the area from the British.

Cadillac chose to build the fort along what is known today as the Detroit River, the strait between Lake Erie and Lake St. Clair. It was named Fort Pontchartrain (PAHN•chuhr•trayn), but sometimes *du Detroit*, which means "on the strait," was added to the name.

By 1707 farming began outside the fort. The farmland was divided so that each farm had land along the water's edge. The farms were narrow near the river, with long fields stretching inland. They were called "ribbon farms" because of their shape.

At first, few French families settled near the fort. In time, however, a town grew up around the fort. After 1751 both the town and the fort came to be known as Detroit.

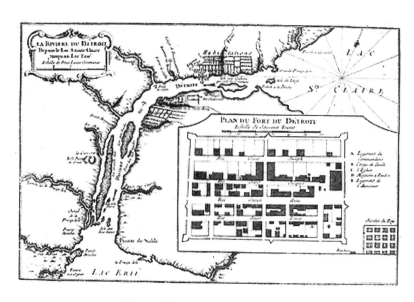

First printed map of Detroit

1. Why do you think the fort was built at a bend in the river? _____

_____

2. What landforms along the river might have blocked the view from the fort?

_____

3. Why do you think it was it important to farmers to be close to the river?

_____

46 MICHIGAN STATE ACTIVITY BOOK UNIT 3

ACTIVITY **9B**

SKILL: CHART AND GRAPH

Fort Pontchartrain

DIRECTIONS: Read the description of Fort Pontchartrain. Then use the word bank to add labels to the diagram below.

Cadillac built Fort Pontchartrain to be sturdy. During construction of the fort, Cadillac ordered his men to cut down trees that were 20 feet (6m) tall and build a palisade, or a fence of stakes. The palisade formed a 200-foot (61-m) square on a bluff above the river. Then they placed a bastion, or tower, at each corner. The bastions stuck out from the fort, serving as lookout points. Around the fort the French built a moat— a deep, wide trench filled with water.

The fort also served as a home for French soldiers and settlers and as a trading post. The widest gate faced the river, so large loads of furs or other goods could be brought in. One main street ran parallel to the river inside the palisade. Other, smaller streets connected to it. Along the main street, workers built small wooden houses and a church. The fort also held a large warehouse with a store for trading inside.

| palisade | bastion | front gate | house |
| church | moat | warehouse/ trade store | main street |

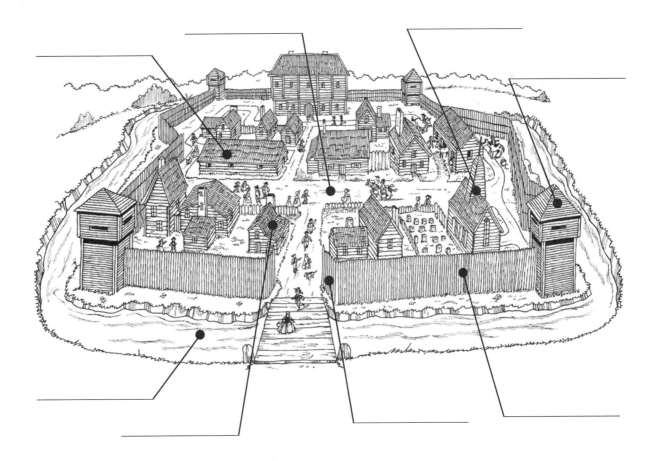

UNIT 3

MICHIGAN STATE ACTIVITY BOOK 47

Lesson 10

THEME: GEOGRAPHY

The Fur Trade

MAIN IDEA: The French fur trade grew in North America because of the high demand for furs in Europe.

In the 1600s beaver fur was **scarce**, or hard to find, in Europe. Yet there was a great demand for the furs. Europeans valued beaver hats as a status symbol, or sign of a high place in society. Because fur products sold for high prices in Europe, the French expanded the fur trade in North America during this time.

Fur-trading centers quickly grew along the lakes and rivers of the Great Lakes region. The Straits of Mackinac, where fur traders could travel from Lake Michigan to Lake Huron, was a center of the French fur trade in North America. Traders also used the St. Marys, the St. Clair, the Detroit, the St. Joseph, the Kalamazoo, and the Grand Rivers. Many French missions and forts along these waterways became busy settlements.

Native Americans traded furs for goods made in Europe. The French offered beads, knives, guns, blankets, shirts, mirrors, and metal pots and tools in trade.

When the French traders gathered a large number of furs they gave them to **voyageurs** (voy•uh•ZHERZ). *Voyageur* is a French word meaning "traveler." The voyageurs took the furs to Montreal, a port city on the St. Lawrence River. From there, the furs were shipped to Europe. While at Montreal, the voyageurs picked up more goods to trade with Native Americans. Voyageurs traveled and lived as the Native Americans did. They also spoke many native languages.

European traders offered metal pots, beads, and other items in exchange for furs.

About 10 million beavers lived in North America when the Europeans arrived. Over time the animals became threatened because of too much hunting. The beaver population increased again starting about 1850, when silk hats became fashionable in Europe and the demand for beaver furs dropped.

RECALL
1. Why did the fur trade grow in North America?
2. What did the Native Americans get in exchange for the furs?

CRITICAL THINKING—DISCUSS OR WRITE
3. **APPLY** How were the rivers and lakes of Michigan important to the fur trade in North America?
4. **ANALYZE** Skins from animals that were larger than a beaver were often worth much less than a beaver skin. Why do you think this was so?
5. **CORE DEMOCRATIC VALUES: DIVERSITY** How might the French have shown respect for Native American cultures?

ACTIVITY 10

SKILL: CHART AND GRAPH

Be a Fur Trader

Today we exchange coins and dollars for goods. When Native Americans traded with the French, they used animal skins instead of money. Prices were set according to an object's value in beaver skins. For example, a blanket might cost four beaver skins, and an ax might cost two beaver skins. This kind of trading of one good or service for another without using money is called **bartering**.

Often trappers would buy hunting gear at trading posts on credit. This meant they paid for the gear later, when they brought back furs.

DIRECTIONS: Study the table at right which shows the different items that could be traded for beaver skins. Then answer the questions.

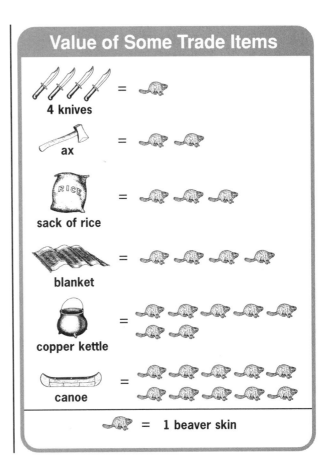

1. How many beaver skins would it take to buy 4 knives and 4 blankets? _____

2. How much would it cost to buy a canoe and an ax? _____

3. You have 32 beaver skins. You need 4 knives, 1 ax, 1 sack of rice, and 4 blankets. How many skins will you have left over? _____

   Will you have enough to buy a canoe? _____

4. How do you think people decided how many furs a blanket was worth?

   _____

   _____

   _____

LESSON 11

THEME: CONFLICT AND COOPERATION

The French and Indian War

MAIN IDEA: France and Britain fought for control of land in North America from 1754 to 1763.

As more Europeans settled in North America, control of the land and its resources became more important to them. Both Britain and France claimed some of the same lands. From 1754 to 1763 these two countries fought in what became known as the French and Indian War. Before the war, nearly all the lands around the Great Lakes were claimed by France. Britain had most of its colonies along the Atlantic Ocean.

During the war Native American tribes were **allies**, or partners, with the Europeans. Often alliances, or relationships between allies, were based on long-standing ties that grew from trading with one another. The Haudenosaunee (hoh•DEE•noh•shoh•nee), or Iroquois (IR•uh•kwoy), people sided with the British. The Ottawas and the Ojibwas sided with the French. One Ottawa chief, named Pontiac, fought for the French because he had benefited from the French fur trade.

No battles were fought in Michigan. Yet tribes from the region fought in a battle in 1755 that was France's biggest victory of the war. Charles Langlade led a group of Ojibwas and Ottawas from Michigan who were sent to the Ohio Valley to defend the French Fort Duquesne (doo•KAYN). Langlade's group helped to defeat the larger British army by using Native American ways of fighting. General Edward Braddock, with a young officer from Virginia named George Washington, led the British forces. Braddock was killed during the battle, and the British retreated.

In 1760 the French had to surrender Fort Pontchartrain to the British. While watching the British flag being raised over the fort, native people told the British, "This country was given by God to the Indians."

In 1763, at the end of the war, France signed the Treaty of Paris. As a result, France lost almost all of its lands in North America. Britain gained control of Canada and almost all the French land east of the Mississippi River.

RECALL
1. What land areas were claimed by France before the French and Indian War?
2. Who won the French and Indian War?

CRITICAL THINKING—DISCUSS OR WRITE
3. **REFLECT** How do you think the conflicts between the French and the British might have been resolved *without* fighting?
4. **EVALUATION** Why do you think Native Americans fought on both sides in the French and Indian War?
5. **INQUIRY** What other questions do you have about the French and Indian War? Use history books or Internet sites to find the answers.

ACTIVITY 11

SKILL: IDENTIFY POINTS OF VIEW

Examine Native American Alliances

DIRECTIONS: Read the following statements to learn why some Native American groups supported the British during the French and Indian War, while others supported the French. Then answer the questions that follow.

British Allies	French Allies
"We are the powerful Iroquois League of Five Nations. We support the British because they offer us more and better goods for fewer furs than the French do. We signed a treaty with the British over control of the Ohio River valley. We will fight to protect our interests in the fur trade. We need the fur trade to survive. We need to expand to new areas where fur is still plentiful."	"We are the Ottawa, Huron (Wyandot), Chippewa (Ojibwa), Potawatomi, and Sac tribes. We also include the Lenni Lenape, who were forced west by the Iroquois. We fear that the Iroquois will take our fur-trading lands in the Ohio River valley, so we joined the French. Fewer French live in the region, so they are less of a threat to our ways of life than the British settlers are."

1. Why did the Iroquois League of Five Nations support the British?

   _____

   _____

   _____

2. Why did the Ottawa, Huron, Chippewa, Potawatomi, and Sac tribes support

   the French? _____

   _____

   _____

3. Imagine that you are a Native American living in the Great Lakes region during the French and Indian War. Whom would you support during the war, the British or the French? Explain your reason.

   _____

   _____

   _____

   _____

LESSON 12　　　　　　　　　　　　　　**THEME: CONFLICT AND COOPERATION**

Pontiac's Rebellion

MAIN IDEA: In 1763 an Ottawa chief named Pontiac led a fight against British control of the Great Lakes region.

As a result of the French and Indian War, the British took over forts in the Great Lakes region from the French. The forts protected British settlers in the area and were important trading centers.

At first, the British and the Native Americans in the Great Lakes region **cooperated**, or agreed to help each other. The British promised more fur trading and fewer settlers. In return, an Ottawa chief named Pontiac promised that British soldiers could pass through Indian territory without being attacked.

The British, however, quickly broke their promises. Soon new settlers began to move onto tribal land, especially in the Ohio valley. The British also stopped Native Americans from trading inside their forts.

The Native Americans were angered. They did not want to lose more of their hunting grounds to British settlers. In 1763 Pontiac brought together the Native American tribes of the Great Lakes region. Over the next few months, Pontiac's fighters rebelled and attacked British forts and settlements. They captured many forts.

In May 1763 Pontiac's fighters tried to capture Fort Detroit but failed. For more than 130 days they surrounded the fort in a **siege**, or an ongoing attack. As winter neared, Pontiac agreed to end the siege. He wrote to the fort's commander, "All my young men have buried their hatchets. I think you will forget the bad things which have taken place for some time past. Likewise, I shall forget what you have done to me, in order to think of nothing but good." In 1766 Pontiac signed a peace treaty with the British.

Chief Pontiac

RECALL
1. Why did the British take over forts along the frontier?
2. Why did Pontiac lead a rebellion? What happened as a result?

CRITICAL THINKING—DISCUSS OR WRITE
3. **SYNTHESIZE** How would promises to cooperate help both the British and the Native Americans?
4. **EVALUATE** How might Pontiac's Rebellion have affected how the British and Native Americans thought about each other? How do you think the events around Pontiac's Rebellion affected the way the groups would cooperate in the future?
5. **CORE DEMOCRATIC VALUES: THE COMMON GOOD** How might the cooperation between the British and the Native Americans be considered an example of a core democratic value? How might the failure of the British to keep their promises be an example of not living up to a core democratic value?

ACTIVITY 12

SKILL: IDENTIFY CAUSE AND EFFECT

The Proclamation of 1763

DIRECTIONS: A *cause* is an event or action that makes something else happen. An *effect* is what happens as a result of that event or action. Read about the causes and effects of the Proclamation of 1763. Then complete the chart, and answer the question that follows.

To avoid more conflicts between colonists and Native Americans after the French and Indian War, the British king issued the Proclamation of 1763. A **proclamation** is an order from a leader to the citizens. The Proclamation of 1763 said that settlers had to leave the territory west of the Appalachian Mountains. Only Native Americans could use those lands. In addition, no one could buy land from the Native Americans. Anyone who wanted to trade with Native Americans would need special permission. The king hoped to make it easier for the British government to control the fur trade in the Great Lakes region by keeping colonists out of the way.

The British found that the proclamation was hard to enforce. Many colonists were angry that the British government was telling them where they could live. In protest, colonists continued to settle in the Indian territories. Their anger at the proclamation was one factor that led to their decision a few years later to declare the colonies independent.

CAUSE
1. The British wanted to prevent more conflicts with Native Americans.

➡

EFFECT/CAUSE
2. _____

⬇

EFFECT/CAUSE
3. After the Proclamation of 1763 was issued, many colonists continued to move west of the Appalachian Mountains.

➡

EFFECT
4. _____

What might have happened as a result of settlers' moving into Indian territories?

_____

_____

_____

UNIT 3

MICHIGAN STATE ACTIVITY BOOK 53

UNIT 3

Practice Test

PART 1 SELECTED RESPONSE

DIRECTIONS: The following is a letter that could have been written by a colonist in what is now Michigan in response to the Proclamation of 1763. Use it and what you have learned in Unit 3 to answer the questions that follow.

> "The Proclamation of 1763 says I must give up my farm and return to the East. I will disobey this unfair law. I have worked hard to clear the land, build a home, and plant my crops near Fort Detroit. The soil here is fertile, and the water is plentiful. It is true that we sometimes have bad relations with the Indians in the area. However, I believe it is Britain's job to protect the colonists here. I am willing to risk settling here. I am unwilling to give up all that I have created here to start again back East. And I believe that the king has no right to force settlers to do so."

1. The Proclamation of 1763 said that settlers could no longer live
   A. north of the Virginia border.
   B. west of the Appalachian Mountains.
   C. along the Hudson River.
   D. in territory controlled by the French.

2. What is the main reason that the colonist does not want to obey the Proclamation?
   A. He does not like being ruled by a king.
   B. He believes the East is poor for farming.
   C. He trades with the Indians.
   D. He does not want to lose his farm.

3. Why did the British issue the Proclamation of 1763?
   A. to prevent conflicts between British colonists and Indians
   B. to make it easier to collect tax money from the colonists
   C. to prevent the French from taking land in Canada
   D. to keep the Mississippi River open for trade

4. Which event helped persuade the British king to issue the Proclamation of 1763?
   A. the arrival of the French in Michigan
   B. the end of the fur trade
   C. the signing of the Treaty of Paris
   D. the start of Pontiac's Rebellion

(continued)

PART 2 EXTENDED RESPONSE PREPARATION

Check Your Social Studies Knowledge

DIRECTIONS: The meeting of European and Native American cultures had a great and lasting effect on Native Americans. Write a statement that tells how Native American life was affected. Support this statement by giving details on how Native American culture was affected by interaction with Europeans. Remember to include both positive and negative effects for Native Americans.

1. How did the arrival of Europeans in North America affect Native Americans?

_____

_____

2. Support your statement:

_____

_____

_____

_____

_____

_____

_____

_____

_____

_____

LESSON 13

THEME: CONFLICT AND COOPERATION

Michigan and the American Revolution

MAIN IDEA: The colonists' desire to govern themselves led them to cut ties with the British government and form their own country.

During the 1760s the British government needed money to pay for the French and Indian War. To raise money, the British added new taxes on items that the colonists used, such as sugar and tea. They also passed new laws for the colonies, including laws that limited where colonists could settle.

These acts angered many colonists. They did not see why they should pay taxes they had not agreed to. After all, they had no **representation** in the British government. This meant that they could not choose leaders to speak or act for them. Throughout the colonies people began to say, "No taxation without representation!"

Many colonists decided to **boycott**, or refuse to buy, British goods. Others responded with violence. They even began to talk about **independence**, or the freedom to govern on one's own. They said that the time had come for the 13 colonies to break away from British rule and start their own country.

Fighting between the colonists and British troops broke out in Massachusetts in 1775. Soon after that, the American Revolution began. A **revolution** is a large, sudden change in government or in people's lives.

Not all the colonists wanted to go to war against Britain. Throughout the colonies people disagreed about which

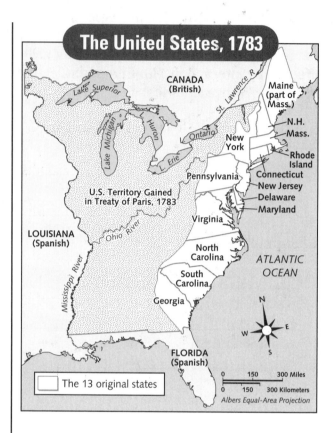

side to support in the American Revolution. Colonists who wanted freedom from Britain were called **Patriots**. Those who still supported Britain were called **Loyalists**. Some colonists remained **neutral** (NOO•truhl), or chose not to take sides. Colonists living in the territories, such as those in the Great Lakes region, had to choose which side to support, too. Most of the colonists in the terrtories chose to become Patriots.

No battles were fought in Michigan, but Fort Detroit played a role in the American Revolution. Since the British controlled it, the fort provided their army with soldiers and supplies. From Fort Detroit the British led their Native

(continued)

American allies in the Great Lakes region in attacks against settlers in Kentucky, Ohio, Pennsylvania, and New York. The Native Americans helped the British because they wanted to keep more settlers from moving onto their hunting grounds in the Great Lakes region. They thought that a British win would prevent colonists from moving west. After all, the British king had promised this land to the Native Americans. Not all the native people helped the British, however. Some fought alongside the Americans. Most stayed neutral.

At first, it seemed unlikely that the Americans would win the fight. Their army had little training. In contrast, the British army was very experienced. Yet as the war continued, the American army grew stronger and stronger.

After many years of fighting, the colonists defeated the British. The Treaty of Paris, signed in 1783, officially ended the Revolutionary War. It also defined the borders of the United States. The new nation stretched from the northern border of what is now Florida to what is now the United States-Canada border. This included the land now known as Michigan. The Atlantic Ocean formed the entire eastern border, and the Mississippi River shaped its western boundary.

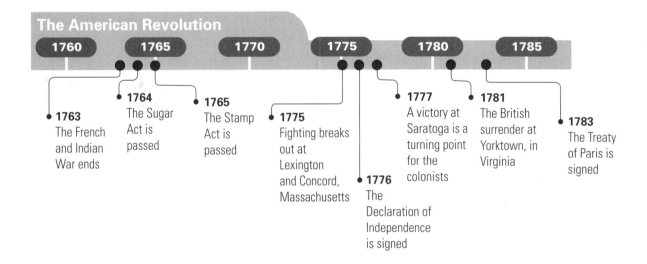

RECALL
1. What role did Fort Detroit play in the American Revolution?
2. How did the United States gain the territory that is now the state of Michigan?

CRITICAL THINKING—DISCUSS OR WRITE
3. **CORE DEMOCRATIC VALUES: POPULAR SOVEREIGNTY** How did the American Revolution help bring about popular sovereignty in the colonies?
4. **EVALUATION** Why did many Native Americans in Michigan support the British? What do you think the Native Americans hoped to preserve by doing so?
5. **INQUIRY** With a partner, find quotations by people on both sides of the Revolutionary War. Explain the point of view that each person is expressing, and compare that point of view with those in other quotations you find.

ACTIVITY 13

SKILL: ANALYZE PRIMARY SOURCES

Examine the Declaration of Independence

DIRECTIONS: Read the paragraphs, and look at the copy of the Declaration of Independence on page 59. Then complete the activities below.

The Declaration of Independence is divided into four parts. The first part, known as the Preamble, tells why it was written. It explains that the colonists felt that they had the right and the duty to form their own government.

The second part of the Declaration describes the colonists' ideas about government. One of the best-known lines comes from this part: "We hold these truths to be self-evident, that all men are created equal, that they are endowed by their Creator with certain unalienable Rights, that among these are Life, Liberty, and the Pursuit of Happiness."

The third and largest part is a list of complaints against Britain's king. It also explains that the colonists tried to settle their differences peacefully. The last part of the Declaration is only one paragraph long. It says that the colonies are free and independent states that are no longer loyal to Britain.

At a meeting known as the Second Continental Congress, representatives from the colonies voted for independence. On July 4, 1776, the group approved the Declaration of Independence. Every year citizens of the United States celebrate July 4 as Independence Day.

1. What are the rights listed in the second part of the Declaration of Independence?

   _____

2. Some signers of the Declaration of Independence came from the wealthiest and oldest families in the colonies. By signing the Declaration, they risked everything their families had achieved. Why do you think they took such a risk?

   _____

3. Each of the following statements is a complaint against the king's actions that is listed in the Declaration. Read each statement and the core democratic value that follows. Then, on a separate sheet, rewrite each statement so that it supports a core democratic value.

   A. "He has refused his Assent to Laws . . ." (The Common Good)

   B. "He has dissolved Representative Houses repeatedly . . ." (Popular Sovereignty)

   C. "He has kept among us, in times of peace, Standing Armies without the Consent of our legislatures." (Liberty)

   D. ". . . For imposing Taxes on us without our Consent" (Popular Sovereignty)

   E. ". . . For depriving us, in many cases, of the benefits of Trial by Jury" (Justice)

(continued)

Preamble

The Preamble tells why the Declaration of Independence was written. It states that the signers believed that the colonies had the right to break away from Britain and become free.

A Statement of Rights

This part of the Declaration tells what rights the signers believed that all people have. All people are equal in having the rights to life, liberty, and the pursuit of happiness. The main purpose of a government is to protect the rights of people who consent to be governed by it. These rights cannot be taken away. When a government tries to take away these rights, the people have the right to change the government or do away with it. The people can then form a new government that respects these rights.

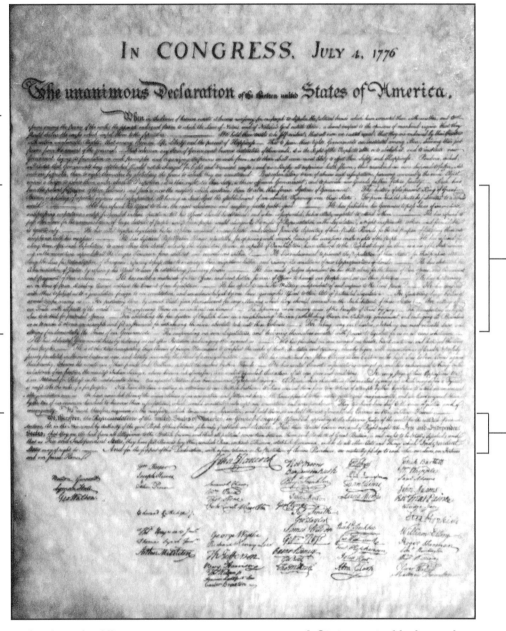

Charges Against the King

The Declaration lists more than 25 charges against the king. It says he mistreated the colonists to gain total control over the colonies. After listing the complaints against the king, the Declaration explains the efforts of the colonists to avoid separation from Britain.

A Statement of Independence

The writers declared that the colonies were now free and independent states. All ties with Britain were broken. As free and independent states, they had the right to make war and peace, to trade, and to do all the things free countries could do.

LESSON 14

THEME: VALUES AND NEW IDEAS

The Constitution of the United States

MAIN IDEA: The Constitution of the United States explains how our government works, and it protects the freedoms of people in the United States.

After the American Revolution, the leaders of the United States had to figure out a plan of government that would unite the states as one nation. In 1787, 55 representatives gathered in Philadelphia to write a plan of government, or a **constitution**. The result was the Constitution of the United States. This document describes the plan for our federal, or national, government.

The Constitution sets up three branches, or parts, of the federal government. The **legislative branch** has the power to make laws. In the federal government, this branch is called Congress. The main duty of the **executive branch** is to see that the laws passed by Congress are carried out. This branch is headed by the President. The main job of the **judicial branch** is to make sure that laws are carried out fairly. The judicial branch includes the most important court in the country—the Supreme Court.

The Constitution gives each branch ways to check, or limit, the power of the other branches. This system is called **checks and balances**. It keeps any one branch from becoming too powerful.

The Constitution also describes the responsibilities that the states, including Michigan, have to the federal government and to each other. All state laws must agree with the Constitution. As one of the 50 states, Michigan cannot make a law that disagrees with the Constitution.

Not long after the first states approved the Constitution, it was changed. Ten **amendments**, or changes, were added to it to protect people's rights. These ten amendments, called the Bill of Rights, describe freedoms that the government cannot take away. They include freedom of religion, speech, assembly, and the press, and the right to a trial by jury. The Bill of Rights also says that the federal government can do only those things that are listed in the Constitution. All other rights and powers belong to the states or to the people.

RECALL
1. What does each branch of the federal government do?
2. How does the Constitution affect Michigan state laws?

CRITICAL THINKING—DISCUSS OR WRITE
3. **ANALYZE** Why do you think it is important that the federal government is separated into three branches? What might happen if our government did not have a system of checks and balances?
4. **CORE DEMOCRATIC VALUES** How does the Bill of Rights show our core democratic values?
5. **EVALUATION** Do citizens have a responsibility to preserve the rights guaranteed to them in the Constitution? Explain why or why not.

ACTIVITY 14

SKILL: MAKE A THOUGHTFUL DECISION

Write a Constitution

DIRECTIONS: Work with a group of classmates to write a constitution for your class. The document may include the rights everyone in the classroom shares. Make sure that everyone in your group agrees on all of the rules in your constitution. Then compare your document with those of other groups in your class. Discuss which parts of your constitution support core democratic values.

Our Classroom Constitution

_____
_____
_____
_____
_____
_____
_____
_____
_____
_____
_____
_____
_____
_____
_____
_____

LESSON 15

THEME: VALUES AND NEW IDEAS

Forming the Northwest Territory

MAIN IDEA: The Northwest Ordinance of 1787 set up a plan for governing the Northwest Territory and made rules for how territories could become states.

After the American Revolution, most of the Great Lakes region became part of the United States. The government named the region the Northwest Territory. It included the unsettled lands west of Pennsylvania, north of the Ohio River, east of the Mississippi River, and south of the Great Lakes.

In 1785 the United States began to divide the Northwest Territory for settlement. However, there was no plan for how the land would be governed. In July 1787 Congress passed the Northwest Ordinance. This **ordinance**, or group of laws, set up a plan of government for the Northwest Territory.

At first, a governor, a secretary, and three judges ruled the vast Northwest Territory. The Northwest Ordinance of 1787 also set up a plan by which the Northwest Territory would be divided into separate, smaller territories that would in time become states. When more than 5,000 adult men lived in a territory, they could form a legislative council and send a nonvoting person to Congress. When 60,000 people lived in a territory, they could ask Congress for permission to write a state constitution and apply for statehood.

Not fewer than 3 states nor more than 5 were to be formed from the Northwest Territory. Each state would have the same powers and rights as the original 13 states.

The ordinance had other important features. Even before the Northwest Territory became separate territories and then states, people who lived there had rights such as freedom of religion. In addition, the ordinance was the first national law that restricted slavery. **Slavery** is the practice of making one person the property of another. The ordinance stated that "There shall be neither slavery nor involuntary servitude in the said territory [the Northwest Territory]." It did, however, allow runaway slaves to be "lawfully reclaimed" and returned to their owners. Finally, the ordinance supported education in the territory. ". . . Schools and the means of education shall forever be encouraged," it said.

(continued)

Michigan was part of the Northwest Territory, which was part of the United States. Even so, British troops occupied Detroit and other posts in Michigan until 1796. In 1798 the first United States election was held. In Michigan, voters elected male residents to serve on the Northwest Territory's legislative council.

In general, pioneers settled first in the eastern part of the Northwest Territory. In time, others moved farther north and west in the territory. At first, few settlers came to Michigan. One reason was that the British supported Native Americans who tried to keep settlers off their lands. Also, traveling conditions to Michigan were poor, and many people thought the land was too wet to be good for farming.

TERRITORY TO STATEHOOD
Set up rules to turn territories into states, based on their growing populations

RIGHTS
Gave rights to people living in territories, based on the Bill of Rights and the Constitution

KEY PARTS OF THE NORTHWEST ORDINANCE

SLAVERY
Outlawed slavery in territories; did not give special rights to runaway slaves

EDUCATION
Encouraged education

RECALL
1. What law set up a plan for governing the Northwest Territory?
2. What present-day states were part of the Northwest Territory?

CRITICAL THINKING—DISCUSS OR WRITE
3. **EVALUATION** What plan did the Northwest Ordinance set up for governing the Northwest Territory? How did it extend the Bill of Rights?
4. **ANALYZE** Why did the Northwest Territory need its own regional government?
5. **REFLECT** What do you think Native Americans living in the Great Lakes region thought of the Northwest Ordinance of 1787? Explain.

ACTIVITY 15   SKILL: MAKE INFERENCES

Townships in the Northwest Territory

DIRECTIONS: When you make inferences, you use facts from your reading as well as what you already know. Read the facts below. Then, on a separate sheet of paper, answer the questions that follow to help you make inferences about dividing land into townships.

Facts About Dividing the Northwest Territory

- To better plan for the settlement of western lands, Congress decided to **survey**, or measure, the Northwest Territory.

- Congress passed the Land Ordinance of 1785, which stated that the Northwest Territory was to be divided into squares called **townships**.

- Each township measured 6 miles (about 10 km) wide by 6 miles long.

- Each township was divided into 36 smaller, square **sections** to be sold to settlers. All the sections were about the same size.

- Each section was numbered. Section 16 of each township was set aside for the support of public schools.

Questions

1. Why do you think townships were squares rather than other shapes?

2. Not all townships in what would become Michigan were exact squares. Why do you think this was so?

3. Why do you think planners wanted all the sections in a township to be about the same size?

4. Townships had names, but sections had numbers. Why do you think this was so?

5. Why do you think one section of every township was set aside for the support of schools?

LESSON 16

THEME: VALUES AND NEW IDEAS

Michigan Becomes a Territory

MAIN IDEA: In 1805 Michigan separated from the Indiana Territory to become the Michigan Territory.

In 1787 the United States organized the western lands it had gained after the American Revolution into the Northwest Territory. In 1800 Congress created the Indiana Territory out of the Northwest Territory. The Indiana Territory included much of what would become the state of Michigan. The rest of Michigan remained part of the Northwest Territory.

Then in 1803, the state of Ohio was carved out of the Northwest Territory. The part of Michigan that had remained part of the Northwest Territory then became part of the Indiana Territory. The people who lived in Detroit felt that the capital of the Indiana Territory, Vincennes (vin•SENZ), was too far away. They wanted their own territory. Three hundred residents of Detroit signed a **petition**, or a signed request for action. The Detroit petition requested that Congress form a separate territory.

On January 11, 1805, President Thomas Jefferson signed the act that formed the Michigan Territory out of the northern part of the Indiana Territory. The Michigan Territory included all of what is now the Lower Peninsula and a part of the Upper Peninsula. Detroit became the capital of the territory. The first governor was William Hull.

On June 11, 1805, the new Michigan Territory faced its first major problem. A fire started in a bakery in Detroit. The fire spread and burned most of Detroit. One observer of the fire wrote that "in less than two hours the whole town was in flames." Fortunately, none of the town's 900 residents were killed.

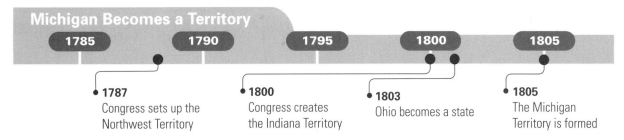

RECALL

1. When was Michigan split from the Indiana Territory to form the Michigan Territory?
2. Why did the people of Detroit want their own territory?

CRITICAL THINKING—DISCUSS OR WRITE

3. **EVALUATION** How did the creation of the Indiana Territory affect settlers in Michigan?
4. **CORE DEMOCRATIC VALUES: POPULAR SOVEREIGNTY** How did the people of Detroit show their belief in self-rule and independence? How might that have shaped their feelings for Michigan's becoming a state in later years?
5. **APPLY** By signing a petition, the people of Detroit showed their point of view on an issue important to them. Why might people in a community today sign a petition that shows their point of view on an important issue that the community faces?

ACTIVITY **16**

SKILL: MAP AND GLOBE

Michigan's Changing Borders

DIRECTIONS: By comparing maps of the same place at different times, you can see how that place has changed over time. Study the maps on page 67 of the Great Lakes region at different times. Then answer the questions below.

1. By 1805, what territories had been formed out of parts of the Indiana Territory and the remaining part of the Northwest Territory? _____

2. How many states had been carved out of the Northwest Territory by 1805? _____ by 1818? _____

3. What feature marked the western border of the Michigan Territory in 1805? _____ in 1818? _____

4. Which covered a larger area, the Indiana Territory in 1800 or the Indiana Territory in 1805? _____

5. How did the Michigan Territory change in size between 1805 and 1818?

6. By 1818, the Michigan Territory included land that had once been a part of what territory in 1805? _____

7. Look at a map of the Great Lakes region today. What present-day states were formed out of the Northwest Territory? _____

(continued)

66 MICHIGAN STATE ACTIVITY BOOK

UNIT 4

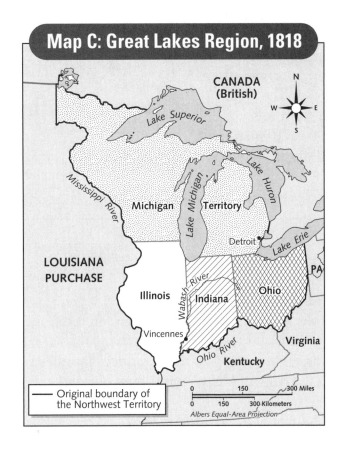

LESSON 17

THEME: HUMAN EXPERIENCES

Pioneer Life in Michigan

MAIN IDEA: Settlers on the American frontier faced many challenges.

In the first half of the 1800s, the whole country seemed to be on the move. Thousands of families **emigrated**, or left their homes, to move west to the Great Lakes region. The territory west of the Appalachian Mountains promised more land and the chance to build a better life. In the Michigan Territory, the population increased from 31,639 to 212,267 between 1830 and 1840.

Life on the **frontier**, or lands beyond settled areas, was hard. There were no houses waiting for pioneer families. Settlers had to build their own homes. To do so, most pioneers chopped down trees from the thick forests of oak and hickory that then covered parts of the Great Lakes region.

After building their cabins, settlers had to clear more trees and bushes before they could plant their crops. Farmers in Michigan found that corn, berries, apples, and vegetables grew well in the area.

Most farms were far from towns. Pioneer families had to be **self-sufficient**, or able to provide for all their needs by themselves. Settlers fished, gathered berries, and hunted deer and other animals. If they had cows, they made butter and cheese from the milk. They ground corn into meal and made it into bread or other dishes. Pioneers collected honey and maple syrup. They made clothes of deerskin and wore moccasins. In winter, pioneer families survived on food such as dried fruits and smoked meats, which they had preserved during the warmer months.

As more families came to Michigan, larger villages were started. They usually were located along rivers and major roads. Villages also grew up near lumber mills and gristmills, or mills that ground corn into cornmeal.

A log cabin on the frontier

RECALL
1. Why did many pioneers come to the Great Lakes region?
2. Why did pioneer families in Michigan need to be self-sufficient?

CRITICAL THINKING—DISCUSS OR WRITE
3. **ANALYZE** In what other ways do you think settlers on the frontier showed themselves to be self-sufficient?
4. **EVALUATE** What kinds of items do you think pioneer families would have had to bring with them to the frontier? Why could they not make these items themselves once they got to the frontier?
5. **REFLECT** Do you think frontiers exist today? Explain.

ACTIVITY 17A

SKILL: ANALYZE PRIMARY SOURCES

Changing Views on Settlement

DIRECTIONS: The primary sources below show opinions about Michigan that were common at the times in which they were written. Read them to learn how opinions about settleing Michigan changed over time. Then, on a separate sheet of paper, complete the activities that follow.

From a letter written by a general stationed at Fort Detroit, 1814:

... it would be to the advantage of Government to remove every inhabitant of the [Michigan] Territory, pay for the improvements, and reduce them to ashes, leaving nothing but the Garrison [military] posts. From my observation, the Territory appears to be not worth defending, and merely a den for Indians and traitors. The banks of the Detroit River are handsome, but nine-tenths of the land in the Territory is unfit for cultivation.

From a Detroit newspaper, 1831:

Emigrant's Song
Come all ye Yankee Farmers,
Who'd like to change your lot,
Who've spunk enough to travel
Beyond your native spot,
And leave behind the village
Where Pa' and Ma' do stay,
Come follow me and settle
In Michigania.

1. Briefly compare the viewpoints about Michigan expressed in the two primary sources. Why do you think viewpoints changed from 1814 to 1831?

2. What does the general's letter tell you about life in the Michigan Territory in 1814?

3. Why might collecting letters and songs be a good way of recording history?

4. If someone wrote a song about Michigan today, how might it differ from the "Emigrant's Song"? How might it be similar?

ACTIVITY **17B**   SKILL: WRITING

Getting to the Frontier

DIRECTIONS: Read about how pioneers traveled to the Great Lakes region. Then imagine that you are a pioneer traveling to the Michigan Territory in the 1800s. On a separate sheet of paper, write a letter about your journey. Be sure to describe each form of transportation that you use.

To get to the Michigan Territory, pioneers had to make a long journey to the frontier. Many pioneer families traveled to the Michigan Territory by boat. In time, canals, new roads, and new inventions helped them make their journeys.

Pioneers from the East often used the Ohio River to travel west. Many made the journey on flatboats. A **flatboat** is a flat-bottomed raft that can carry large amounts of cargo.

Soon keelboats joined flatboats on the rivers. **Keelboats** were similar to flatboats, but they were pointed at both ends and sometimes used sails. Unlike flatboats, they could travel upstream, against a river's current, by being poled or sailed. However, poling, or pushing, upriver was slow, hard work. It could take more than three months to travel north on a keelboat from New Orleans, Louisiana, to the Great Lakes region.

Starting in the 1820s, many travelers used a new invention, the steamboat. A **steamboat** is a boat powered by a steam engine that turns a large paddle wheel, causing the steamboat to move. These boats had enough power to travel

Steamboat

upstream. From Buffalo, New York, travelers going to the Michigan Territory took a three-day steamboat ride across Lake Erie to Detroit.

The journey to the frontier got easier in 1825 when the Erie Canal opened. The canal connected the Hudson River to Lake Erie. Pioneers coming from New York used the canal as a direct water route to the Great Lakes region. The canal also allowed goods to be shipped cheaply to settlers in the West.

After pioneers reached the Michigan Territory by boat, they then traveled inland by wagon. In 1820 new roads connected Detroit to Pontiac and Mount Clemens. People could also travel by road from Detroit to Saginaw and Port Huron. Most roads, however, were muddy, had few bridges, and were not well kept.

Pioneer families had to take with them everything they would need for their new life on the frontier. There would be no stores where they were going. In their wagon or boat, a family made sure to pack an ax and other tools, a rifle, and an iron plow blade. Families also packed clothing, seeds, and an iron pot.

Flatboat

LESSON 18

THEME: CONFLICT AND COOPERATION

The War of 1812

MAIN IDEA: In 1812 the United States declared war on Britain. It did so in part because the British supported Native Americans in trying to stop American settlers from moving onto western lands.

As more American settlers moved into the Great Lakes region, they pushed the frontier farther west. Sometimes Native Americans tried to stop settlers from taking more land.

The Native Americans were helped in their struggle against American settlers by the British in Canada. The British sold the Indians guns and encouraged them to fight the Americans. Before long, troubles in the western lands helped to push the United States into a second war with Britain—the War of 1812.

In preparing for war, the United States made plans to invade and capture British Canada. In July 1812 Michigan's territorial governor, William Hull, led an army into Canada, but Britain's smaller army defeated it.

That same month the British captured Fort Mackinac. The British surprised the fort's commander, who had not yet heard about the war. The Americans surrendered the fort without firing a shot. On August 16 the British also easily captured Fort Detroit. Governor Hull thought that the Americans would be outnumbered by British forces, so he surrendered the fort quickly. The *New York Evening Post* stated: "[W]e did not expect so deep a stain upon our country's character."

In September 1813 the United States recaptured Fort Detroit. Fort Mackinac, however, remained under British control until the war ended.

On December 24, 1814, the British and the Americans signed a treaty ending the war. Neither side clearly won. Under the terms of the treaty, Britain kept Canada, and the United States kept its territory on the frontier.

Governor Hull surrendered Fort Detroit to the British in August 1812.

RECALL
1. What happened to Fort Mackinac during the War of 1812?
2. When was the peace treaty signed to end the War of 1812?

CRITICAL THINKING—DISCUSS OR WRITE
3. **ANALYZE** Why do you think settlers in the Great Lakes region supported the idea of going to war against Britain?
4. **EVALUATION** Why do you think the British wanted to control Fort Mackinac and Fort Detroit?
5. **REFLECT** What might have happened if Governor Hull had not surrendered Fort Detroit to the British?

ACTIVITY 18

SKILL: IDENTIFY CAUSE AND EFFECT

Causes and Effects of the War of 1812

DIRECTIONS: Use what you have learned about the War of 1812 to complete the cause-and-effect chart below.

CAUSE	EFFECT
_____	The United States declared war on Britain.
Governor Hull knew that the British attacking Fort Detroit would have the help of many Native American fighters.	_____
The British and the Americans signed a peace treaty on December 24, 1814. Neither the United States nor Britain felt that its demands were fully met in the treaty.	_____

72 MICHIGAN STATE ACTIVITY BOOK

UNIT 4

LESSON 19

THEME: CONFLICT AND COOPERATION

Statehood for Michigan

MAIN IDEA: The leaders of Michigan had to solve a number of problems before Michigan could become a state.

In the early 1800s many settlers came to the Great Lakes region, but few came to Michigan. After the War of 1812, the new governor of the Michigan Territory, Lewis Cass, tried to make the territory more appealing to settlers. He built public schools, roads, and lighthouses. He also set up counties. In addition, he persuaded Native Americans in Michigan to move farther west.

Cass's work paid off. In the 1830s the territory's population grew from 27,278 to more than 212,000. Many people began thinking it was time for the territory to become a state. Soon, its leaders were planning a new state government. The first state constitution was approved by Michigan's leaders on October 5, 1835. The United States Congress did not approve Michigan's constitution, however. Two problems had to be solved first.

One problem had to do with land claims. Both the Michigan Territory and the state of Ohio claimed a strip of land west of Lake Erie. This land was known as the Toledo Strip. Unless Michigan gave up its claim to this land, Ohio did not want its neighbor to become a state. Finally, Michigan agreed to give up the Toledo Strip. In return, Congress gave Michigan all of the Upper Peninsula that it did not already own. The *Detroit Free Press* described the opinion of many Michiganians when it called the Upper Peninsula "a region of perpetual [constant] snows."

The other problem was about slavery. In the 1830s Congress and the nation were divided over slavery. Michigan's leaders wanted their state to become a free state. Many Southern states did not want the nation to have more free states than it had slave states. A compromise was finally reached. A **compromise** is a settlement of a disagreement, with each side giving up some of what it wants. Arkansas was admitted as a slave state in 1836. Shortly afterward, in January 1837, Michigan joined the Union as a free state. It became the twenty-sixth state of the United States.

Michigan's leaders voted in 1847 to move the state capital from Detroit to Lansing.

RECALL
1. How did Governor Cass attract settlers to the Michigan Territory?
2. Why did Congress not approve Michigan's first state constitution in 1835?

CRITICAL THINKING—DISCUSS OR WRITE
3. **EVALUATION** Why do you think territorial governor William Cass built roads and lighthouses in the Michigan Territory? What happened as a result?
4. **CORE DEMOCRATIC VALUES: THE COMMON GOOD** Why do you think Michigan's leaders agreed to the compromise over the Toledo Strip?
5. **REFLECT** What might have happened if Michigan had received the Toledo Strip instead of the Upper Peninsula as part of the compromise?

ACTIVITY **19A**　　　　　　　　　　　　　　　　　SKILL: SOLVE A PROBLEM

The Toledo War

The Michigan Territory and its neighbor, the state of Ohio, argued over which of them owned a piece of land called the Toledo Strip. The Toledo Strip included an area of about 500 square miles (1,295 sq km) located near the mouth of the Maumee River. The river formed part of a trade route between Lake Erie and the Gulf of Mexico.

In 1835 both sides sent soldiers to the Toledo Strip. Even though no real fighting occurred, it was called the Toledo War. One soldier was injured, however. Finally, the United States Congress persuaded Michigan to give up the Toledo Strip in exchange for the western part of the Upper Peninsula and statehood.

DIRECTIONS: Work in groups to role-play the debate between the Michigan Territory and Ohio over the Toledo Strip. Fill in the steps in the problem-solving chart below to come up with a solution. Make sure everyone in your group agrees on the solution. The solution should support core democratic values, such as justice. Share your chart with other groups.

1. Identify the problem.	
2. Gather information on the problem.	
3. List three possible solutions.	
4. Think about the advantages and disadvantages of each solution.	
5. Choose the best solution, and form a plan to put it into action.	
6. Think about how well the solution worked. Try to refer to Core Democratic Values in your evaluation.	

ACTIVITY **19B**

SKILL: CHART AND GRAPH

Michigan State Government

Michigan was made the twenty-sixth state in 1837. Leaders of the new state wrote a state constitution that outlined a plan of government for it. Like the United States Constitution, Michigan's constitution separated the government into three branches. By dividing the power of the government among three branches, Michigan's leaders made sure that one individual or group would never be in complete control of the state.

The legislative branch of Michigan's state government makes laws. The Michigan Legislature is set up like the United States Congress, with a House of Representatives and a Senate. The Michigan House of Representatives has 110 members. The Michigan Senate is made up of 38 state senators.

The governor leads the executive branch, which is in charge of carrying out the laws. He or she presents the budget for the state to the legislature. A **budget** is a plan that tells where all of the state's money is to be spent. The governor also looks at bills that have been passed by the legislature and decides whether they should become laws and how they should be carried out.

The judicial branch is in charge of making sure that Michigan laws are fair. The Michigan Supreme Court is the highest court in the judicial branch. Seven members serve on this court.

DIRECTIONS: Complete the chart below, which shows the structure of the Michigan state government. Describe the duties of each branch. Then, on a separate sheet of paper, explain why you think the Michigan state government was modeled after the United States government.

MICHIGAN STATE GOVERNMENT

Legislative Branch	Executive Branch	Judicial Branch

Practice Test

PART 1 SELECTED RESPONSE

DIRECTIONS: The paragraph below is based on an excerpt from *Common Sense*, a pamphlet written by Thomas Paine at the time of the American Revolution. The document helped persuade many colonists to revolt against Britain. After you have read the paragraph, use it and what you already know to answer the questions that follow.

> "I challenge the warmest supporter of British rule to show a single advantage of reuniting with Britain. I can see not a single advantage to doing so. Without Britain's help, we can sell our products in any market in Europe. We must pay for any goods we buy, no matter whether we are ruled by Britain or are independent. But the injuries we receive by our connection with Britain are many. It is our duty to humankind at large as well as to ourselves to remain independent."

1. Which of the following was NOT one of the "injuries" referred to by Thomas Paine?
   A. taxation without representation
   B. forced housing of British soldiers
   C. compromise with the king
   D. British control of colonial trade

2. What advantage would separation from Britain offer to Michigan?
   A. Colonists would be free to trade with Native Americans in Michigan.
   B. Colonists would be free to buy Michigan's farm products.
   C. Colonists would be free to ship manufactured goods to Michigan.
   D. Colonists would be free to travel to and settle in Michigan.

3. In which document did colonists cut all ties with Britain?
   A. the Proclamation of 1763
   B. the Treaty of Paris
   C. the Loyalist Pact
   D. the Declaration of Independence

4. Which core democratic value is Paine asking colonists to support?
   A. equality
   B. popular Sovereignty
   C. life
   D. pursuit of Happiness

(continued)

PART 2 EXTENDED RESPONSE PREPARATION

Use Core Democratic Values

DIRECTIONS: Read the paragraph in the box below. Then answer the questions that follow.

> In 1773 Britain passed the Tea Act. This law said that colonists could buy tea only from a British company called the East India Company. Colonists would also have to pay a tax to Britain on the tea. However, they would have no say in how the British government spent the taxes it collected.

1. If you had been a colonist in 1773, would you have been for or against the Tea Act?

   _____

2. Use core democratic values to support your answer.

   _____
   _____
   _____
   _____
   _____
   _____
   _____
   _____
   _____
   _____
   _____
   _____

Lesson 20

THEME: GEOGRAPHY

Growth as a State

MAIN IDEA: In the first years after Michigan became a state, its population and its economy grew.

Michigan grew quickly in its early years as a state. In the mid-1800s Michigan's population increased rapidly. Thousands of people came to the state to farm and to work in the fast-growing lumber and mining industries.

In the 1830s tens of thousands of settlers moved to Michigan from other states and countries. They came from as far away as Germany, Finland, the Netherlands, Belgium, Britain, Ireland, and Sweden. The people from New England and New York who moved to Michigan were called Yankees. One writer described Michigan's Yankees as "a thrifty, enterprising [ready to take risks], plucky [brave] people, with high ideals of religion, morality, and education."

At first, the people who came to the Middle West to farm did not have much success growing crops in large amounts. The iron plows used in the East were not very good at turning over the thick, tough sod that covered much of the Lower Peninsula. **Sod** is soil held together by grass and its roots. The inventions of two people in nearby Illinois helped farmers in the Middle West. These inventions made farming an important industry in Michigan and in the rest of the Middle West.

John Deere invented a plow made of steel, which could cut right through sod without getting stuck. He set up a factory in Illinois and soon sold thousands of plows a year.

When it was time to harvest grain, farmers faced another hard job. They had to cut the grain by hand, using a tool called a cradle. With a cradle a farmer could cut only one or two acres of grain in a full day. Cyrus McCormick changed that by inventing a mechanical reaper. Using this reaper, which was pulled by horses, a farmer could harvest five times as much as could be done by hand.

In the 1840s lumber and mineral resources in the Upper Peninsula also attracted people to Michigan. Mine owners in the Upper Peninsula encouraged people to come there to mine iron ore and copper. People also flocked to Michigan to

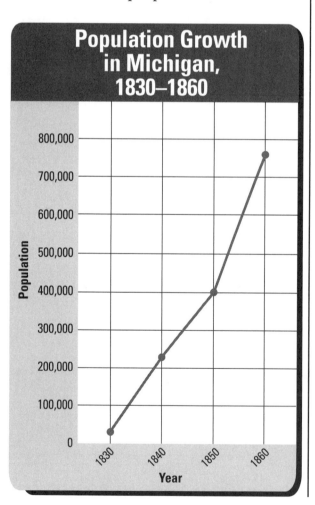

(continued)

work for lumber companies that cut logs into boards ready for use. The dense forests of the state supported a fast-growing lumber industry. By 1840 Michigan had more than 500 sawmills.

Another important reason for Michigan's growth was improvements in transportation. To transport, or move, lumber and other products, Michigan built more roads, canals, and railroads. These improvements in transportation helped businesses grow in Michigan and brought more people to the state. The opening of the Soo Canals in 1855 was especially important to the mining industry in Michigan. Mining companies could now ship copper and iron ore directly from the mines in the Upper Peninsula to factories in Detroit, Cleveland, and other cities along the Great Lakes. The Erie Canal was also important to the growth of Michigan. The canal, which connected Lake Erie to the Hudson River, helped people and goods get to and from Michigan.

McCormick's mechanical reaper

RECALL
1. What inventions helped places in the Middle West become important farming areas?
2. Where were Michigan's copper and iron ore mines?

CRITICAL THINKING—DISCUSS OR WRITE
3. **REFLECT** Suppose you live in Michigan in the 1840s. What would you do for a living? Why?
4. **EVALUATION** With improved ways of farming, farmers in the Middle West were able to grow more crops than they needed for their own use. How do you think this helped farmers? Why might prices of a crop such as corn drop if there was too much of it for sale?
5. **ANALYZE** How did improvements in transportation help Michigan's economic growth? Give examples.

ACTIVITY **20A**

SKILL: MAP AND GLOBE

Mining in Michigan

DIRECTIONS: Read the paragraphs. Then complete the map by following the steps.

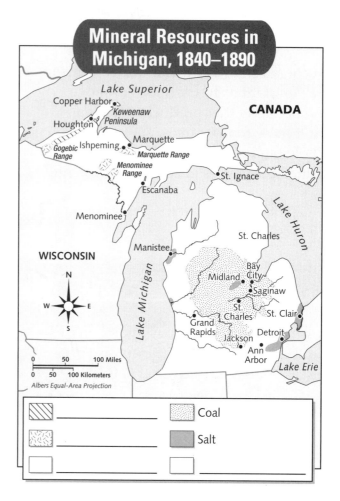

Mining has been a major industry in Michigan since the 1800s. In 1844 an Ojibwa chief led members of a mining company to a rich source of iron ore near Teal Lake. Workers broke up rock from deep beneath Earth's surface to get the iron ore. In time, new technology—dynamite and power drills—made mining the iron ore easier. By 1890 Michigan produced more than three-fourths of the United States's iron ore. Sources of iron included the Gogebic (goh•GEE•bik) Range, the Marquette Range, and the Menominee Range. Today the Marquette Range produces about one-fourth of the nation's iron ore.

In the 1840s a copper boom took place in the northwestern Upper Peninsula. From 1847 to 1887, Michigan produced more copper than any other state. Getting to the copper was hard, expensive, and dangerous work. Miners had to drill deep under the ground and blast the rock with dynamite.

Michigan has other mineral resources besides iron ore and copper. In 1881 a miner found gold near Ishpeming (ISH•puh•ming) in Marquette County. In the 1890s coal mines operated near Jackson, Bay City, and St. Charles. The Dow Chemical Company mined bromine (BROH•meen) near Midland. **Bromine** is a mineral that is used to treat salt water and to make chemicals that put out fires. In 1910 a large salt mine opened in Detroit.

1. Write the words *copper* and *iron* next to the correct patterns in the map key.

2. On the map, circle the area that still produces iron.

3. Underline the names of cities that likely were important salt-mining centers in Michigan.

4. Place a symbol for gold near where that resource was found in Michigan in 1881. Add the symbol to the map key.

5. Place a symbol for bromine near where a chemical factory mined that resource in the 1890s. Add the symbol to the map key.

ACTIVITY **20B**

SKILL: DRAW CONCLUSIONS

Lumbering in Michigan

DIRECTIONS: Read the paragraphs below. Then fill in the chart that follows to draw conclusions about lumbering in Michigan in the 1800s.

By the 1840s Maine and New York had cut down most of their white pine trees. Loggers found new sources of this valuable natural resource in the Lower Peninsula of Michigan, especially in the Saginaw Valley. Many of Michigan's beautiful white pine trees were 300 years old and up to 200 feet (61 m) tall.

Workers called "shanty boys" cut down the trees in the winter. Horses dragged the trees on sleds to frozen waterways. In the spring, when the rivers thawed, lumber workers floated the trees down rivers to sawmills. There they were cut into lumber. Sawmills then shipped the cut lumber on rivers to ports on the Great Lakes. From the Great Lakes, the lumber was shipped to markets in the East.

Lumber production in Michigan increased in the 1870s when narrow railroad tracks were built through Michigan's forests. Trains allowed logging to continue year-round and in places far from rivers and creeks. Loggers in the 1800s usually did not replant the trees. This led to **deforestation**, or the clearing of forestland by cutting down trees. Deforestation caused

Logs piled on sled

problems. The stumps and branches created fuel for forest fires that raged for many years.

People began to replant Michigan's forests. Today, thousands of acres of forests can be found in Michigan. They are owned by private citizens, the state of Michigan, and even the United States government. Since the 1930s people have worked hard to protect the future of Michigan's forests.

FACTS I KNOW		NEW FACTS		CONCLUSION
	+	Narrow railroad tracks allowed logging year-round and far from rivers and creeks.	=	
	+	The state owns some of the land deforested in the 1800s and has replanted trees.	=	

UNIT 5

MICHIGAN STATE ACTIVITY BOOK 81

LESSON 21

THEME: CONFLICT AND COOPERATION

Slavery and Freedom

MAIN IDEA: Many enslaved people tried to gain freedom by running away to northern states, such as Michigan.

Slavery in North America was an issue that had divided Americans since colonial times. By the early 1800s most northern states did not allow slavery. Slavery was allowed in the southern states, however. Many enslaved people ran away from their owners. These runaways were called **fugitives** (FYOO•juh•tivz). After crossing the Ohio River, many fugitives made their way to Michigan, which was a free state.

In the mid-1800s some people helped fugitives reach free states or Canada on the Underground Railroad. The **Underground Railroad** had no train. It was a system of escape routes leading to places that did not allow slavery.

"Stations," or hiding places, along the Underground Railroad included homes barns, and churches. At these stations, slaves received food and blankets and directions to the next safe place. Michigan had a station every 15 miles (24 km). Detroit was one of the last stations before Canada. People called the Detroit River "Route One."

Even though slavery was not allowed in Michigan, it was against the law to protect

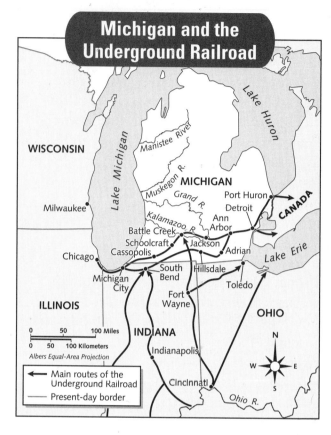

slaves who had escaped from other states. Slave catchers came to Michigan to try to recapture slaves for a reward. Although helping fugitives meant breaking the law abolitionists often chased slave catchers away. **Abolitionists** were people who were against slavery and wanted to abolish, or end, slavery in the United States.

RECALL
1. What was the Underground Railroad?
2. Who were abolitionists?

CRITICAL THINKING—DISCUSS OR WRITE
3. **EVALUATION** Why do you think the Detroit River was sometimes called "Route One"? Why do you think a runaway slave would try to reach Detroit?
4. **CORE DEMOCRATIC VALUES: LIBERTY** How did abolitionists show that they supported the core democratic value of liberty?
5. **SYNTHESIZE** Why do you think people involved in the Underground Railroad took risks to help others?

ACTIVITY 21

SKILL: CATEGORIZE

Sojourner Truth

DIRECTIONS: Read about Sojourner Truth, an abolitionist from Michigan. Then use facts from her life to describe how Sojourner Truth showed the character traits listed below.

Isabella Van Wagener was born into a slave family in Ulster County, New York. As a slave she had five owners before being set free in 1827. A year later, New York made slavery against the law, and the rest of her family was set free.

Van Wagener had always been a deeply religious person. In 1843, at the age of 46, she took the name Sojourner Truth and began to preach her message of love and concern for others. Soon she also became an abolitionist. Sojourner Truth traveled throughout the northern United States speaking out against slavery. She often attracted large crowds. By 1850 she had also begun to fight for women's rights. She used her talents as a speaker to prove that men and women were equal. In one of her most famous speeches, "Ain't I a Woman?," Truth argued that as a slave she worked alongside men, doing the same kinds of jobs. She pointed out that she "ploughed and planted, and gathered" and, besides that, gave birth to 13 children.

Later in her life Sojourner Truth worked to help freed slaves find jobs and homes. In the 1870s she came up with a plan to use lands in the West as farms for former slaves. On November 26, 1883, Sojourner Truth died in Battle Creek, Michigan. She was 86 years old.

1. COURAGE: _____

_____

_____

2. CIVIC VIRTUE: _____

_____

_____

3. COMPASSION: _____

_____

_____

4. PERSEVERANCE: _____

_____

LESSON 22

THEME: CONFLICT AND COOPERATION

Michigan in the Civil War

MAIN IDEA: Michigan soldiers fought in every major battle of the Civil War and helped the Union keep the country united.

In 1860 Abraham Lincoln was elected President of the United States. Lincoln opposed the spread of slavery to new states joining the Union. Many people in the South feared he would try to end slavery everywhere.

After Lincoln's election, 11 Southern states decided to **secede** (sih•SEED) from, or leave, the Union. These states formed their own country, called the Confederate States of America, or the Confederacy. The United States became divided into two parts. People in Missouri, Kentucky, Maryland, and Delaware were torn between the two sides. These **border states** were between the North and the South. Although these states allowed slavery, they remained part of the Union.

When it came time for people in the border states to choose sides, some selected the North. Others sided with the South.

On April 12, 1861, Confederate soldiers fired shots at Union forces at Fort Sumter, in South Carolina. This event marked the beginning of the American Civil War. A **civil war** is a war between people of the same country. For the next four years, the Union and the Confederacy fought the bloodiest war in United States history.

Michigan helped the Union fight the Civil War in many ways. Michigan was the first western state to send volunteers to fight in the war. On May 11, 1861, members of the First Michigan Volunteer Infantry gathered in Detroit. They received their battle flag and left two days later for Washington, D.C., to go fight in the war. When the Michiganians

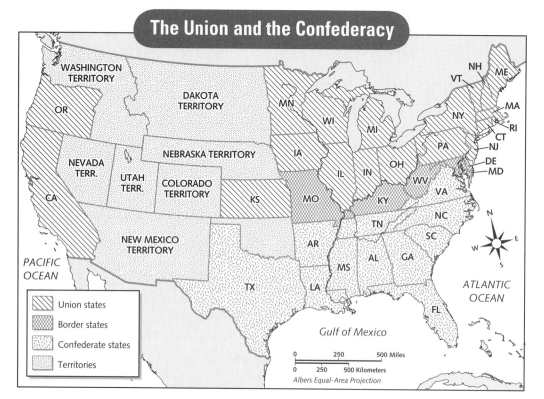

(continued)

84 MICHIGAN STATE ACTIVITY BOOK · UNIT 5

arrived in the capital, President Lincoln greeted them by saying, "Thank God for Michigan."

By the end of the war, about 90,000 men and some women from Michigan had served in the Union armies. Michigan soldiers fought in every major battle of the war, including First Bull Run, Antietam, and Gettysburg. Michigan also had many war heroes. At Gettysburg, General George Armstrong Custer, from Monroe, made a name for himself. At one point during the battle, Custer placed himself in front of his men. He ordered them to charge at the approaching Confederate cavalry, even though they were outnumbered. The battle was difficult, but Custer's men fought bravely, and the Confederates retreated. Michigan's soldiers also included a group of 1,500 African Americans who served with the First Michigan Colored Infantry.

In 1865 the Union won the Civil War. The United States stayed together as a nation, but the price was high. More than 600,000 soldiers had died, more than in any other United States war. About soldiers from Michigan, General Orlando Willcox of Detroit said, ". . . we have done no more than that duty which every citizen owes to a free . . . government."

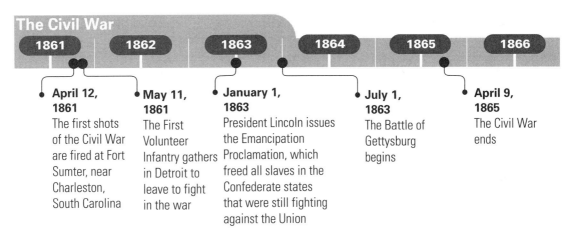

The Civil War

- **April 12, 1861** The first shots of the Civil War are fired at Fort Sumter, near Charleston, South Carolina
- **May 11, 1861** The First Volunteer Infantry gathers in Detroit to leave to fight in the war
- **January 1, 1863** President Lincoln issues the Emancipation Proclamation, which freed all slaves in the Confederate states that were still fighting against the Union
- **July 1, 1863** The Battle of Gettysburg begins
- **April 9, 1865** The Civil War ends

RECALL
1. Why did the Southern states secede from the Union?
2. How did people from Michigan help the Union fight the Civil War?
3. Who won the Civil War?

CRITICAL THINKING—DISCUSS OR WRITE
4. **CORE DEMOCRATIC VALUE: PATRIOTISM** What did General Willcox mean when he said, ". . . we have done no more than that duty which every citizen owes to a free . . . government"?
5. **SYNTHESIZE** Ships carrying supplies for Union soldiers traveled in and out of Michigan ports. Predict what might have happened if the Confederacy had gained control of these ports.
6. **INQUIRY** With a partner, write questions about how Michigan soldiers played roles in major Civil War battles, such as First Bull Run, Antietam, or Gettysburg. Research the answers to your questions, and present the information to the class in an oral report.

ACTIVITY **22A**

SKILL: SEQUENCE

The Battle of Gettysburg

Although no Civil War battles were fought in Michigan, the state's soldiers played an important role at the Battle of Gettysburg, the biggest battle of the war. At Gettysburg, in Pennsylvania, Union soldiers were outnumbered by Confederate soldiers. With the help of thousands of soldiers from Michigan, however, the Union army finally defeated the Confederates.

DIRECTIONS: Place the letter of each event related to the Battle of Gettysburg in its correct place on the time line. Then answer the questions that follow.

A	By the afternoon of July 1, 1863, the Twenty-fourth Michigan Infantry retreats. Only about 100 men out of 496 survive.
B	On the third day of the battle, Union General George A. Custer leads four Michigan Cavalry **regiments**, or military units, to victory.
C	Colonel Jeffords of the Fourth Michigan Infantry bravely tries to save his unit's flag on July 2, 1863. He receives a deadly wound.
D	On the morning of July 1, 1863, the Twenty-fourth Michigan Infantry fights against the better-armed Confederate army in Gettysburg. The unit's stand allows time for other Union troops to arrive.
E	On July 4, 1863, Confederate General Robert E. Lee retreats, marking the turning point of the war in favor of the Union side. Out of almost 4,000 Michigan soldiers, one third are dead, injured, or missing.

The Battle of Gettysburg

July 1, 1863 — July 2, 1863 — July 3, 1863 — July 4, 1863 — July 5, 1863

1. What period of time does this time line cover?

   _____

2. Which event came first, the Twenty-fourth Michigan Infantry's retreat or the Michigan Cavalry's victory?

   _____

   _____

3. Did the colonel of the Fourth Michigan Infantry die during or after the first day of battle? _____

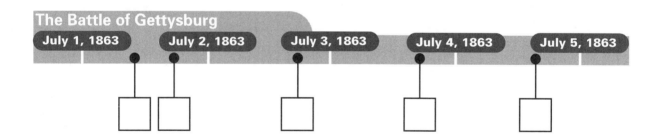

86 MICHIGAN STATE ACTIVITY BOOK UNIT 5

ACTIVITY **22B**

SKILL: GENERALIZE

Women in the Civil War

DIRECTIONS: Read the paragraphs below. Look for facts about the role of Michigan women in the Civil War. Then fill in the chart to make a generalization.

At the time of the Civil War, women were not allowed to serve in the military. During the war most Michigan women stayed home when their husbands, brothers, and sons went off to fight. The women ran farms, managed businesses, and started "relief societies." Relief societies sent boxes and barrels of goods such as newspapers, socks, pins, thread, underwear, jam, and pickles to soldiers. They also operated hospitals near battlefields. Michigan women sent money, clothing, and hospital items, such as bandages, to the Michigan Soldiers Relief Association in Washington, D.C. At the Michigan State Fair in 1864, women sold farm animals, farm products, and other valuable goods. The sales raised $9,000 for the soldiers.

Even though they were not allowed to, a few Michigan women did fight in Civil War battles. In 1861 Sarah E. Edmonds of Flint dressed as a man to join the army. She served as a nurse for two years under the name Franklin Thompson. Later she worked as a Union spy. She left the army when she got sick. She feared that her identity would be discovered at the hospital.

Sarah E. Edmonds

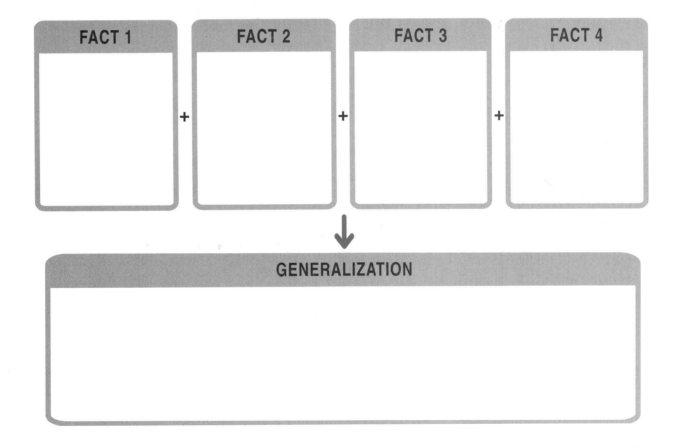

UNIT 5

MICHIGAN STATE ACTIVITY BOOK 87

LESSON 23

THEME: INTERACTIONS

Cities and Industries Grow in Michigan

MAIN IDEA: After the Civil War, cities grew in Michigan. Detroit and other cities became manufacturing centers.

In the years after the Civil War, Michigan changed in many ways. The Civil War had created new opportunities for manufacturing in Michigan. Following the war Michigan's economy continued to become more and more industrial. An **industrial economy** is an economy in which factories and machines manufacture most goods.

Before the war much of Michigan's economy was based on the sale of natural resources—specifically copper, iron, and lumber. After the war the economy depended mainly on making products from these natural resources. Most manufacturing was done in Detroit and other large cities. As businesses in these cities grew, there was a need to keep track of them. In 1870 Ralph Lane Polk began to record information about businesses in business directories or guides. His first directory listed more than 600 businesses —from shoe makers to ship builders—in Detroit alone.

With iron ore from the Upper Peninsula, factories in Michigan cities began making ships, railroad cars, iron stoves, steam engines, and farming tools. In the 1890s, the Michigan Stove Company, which was the world's largest makers of stoves employed more than 1,200 people. It made more than 76,000 Garland stoves each year. Factories in Detroit, such as the Detroit Car and Manufacturing Company and the Michigan Car Company, produced most of the nation's railroad cars. In 1871 George Pullman opened a plant that manufactured his invention, the Pullman sleeping car. This car changed the way people traveled by train, making it more comfortable and private. In addition to railroad cars, Detroit factories were making parts for trains, such as wheels and axles. In 1868 Detroiter William Davis perfected the refrigerated rail car. One year later, the first shipment by refrigerated car was made from Detroit to Boston.

Other new products being made in Michigan included chemicals, cement, and pharmaceuticals (far•muh•SOO•tih•kuhls). **Pharmaceuticals** are medicines. In addition, in the late 1800s furniture made from

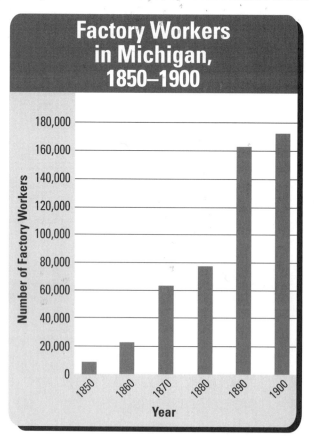

(continued)

Michigan lumber became famous across the country. Much of it was manufactured in factories in Grand Rapids.

In the Lower Peninsula, farmers sold their goods to flour mills, cheese factories, and other kinds of food-processing plants. **Food processing** is the cooking, canning, drying, or freezing of food and the preparing of it for market. During the late 1800s, food-processing plants in Battle Creek began making breakfast cereals.

The many new factories in Michigan needed a lot of workers. Some companies placed job advertisements overseas. These advertisements attracted immigrants from Europe and some from Southwest Asia. By 1920 one-fifth of all the people in Michigan had been born in another country. People also came to Michigan from other parts of the United States, especially the South. Many African Americans from the South moved to the state looking for more equality and better jobs. Some hoped to turn their services and skills into small businesses, such as tailor and carpentry shops.

As more people came to Michigan for factory jobs, cities in the state grew. By 1900 Detroit had 285,000 people. It was the eighteenth-largest city in the United States. About 90,000 people lived in Grand Rapids, the state's next-largest city. Saginaw and Bay City each had about 40,000 residents. By 1920 more people in Michigan lived in cities than in rural areas. Detroit became the fourth-largest city in the United States, with 993,678 residents.

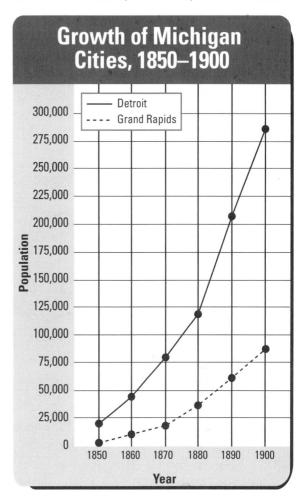

RECALL
1. How did the economy of Michigan change after the Civil War?
2. By 1920 where did most people in Michigan live—in urban areas or rural areas?
3. By how much did the population of Detroit increase from 1900 to 1920?

CRITICAL THINKING—DISCUSS OR WRITE
4. **ANALYZE** How did the growth of manufacturing in Michigan lead to the growth of the state's urban areas?
5. **EVALUATION** Why do you think immigrants were willing to leave their homelands for work in Detroit? How do you think their lives changed?
6. **INQUIRY** In small groups, do research to find out what some of the immediate and long-term effects of the growth of industries in Michigan were. Present your findings to the class.

ACTIVITY **23A**

SKILL: SEQUENCE

Building an Automobile Industry in Michigan

DIRECTIONS: Study the diagram, and read the paragraphs below. Then answer the questions that follow.

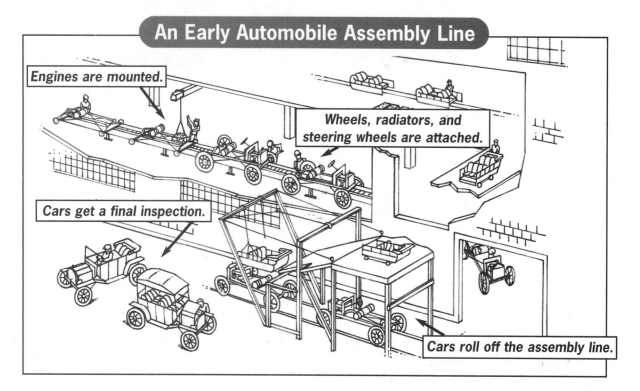

An Early Automobile Assembly Line
- Engines are mounted.
- Wheels, radiators, and steering wheels are attached.
- Cars get a final inspection.
- Cars roll off the assembly line.

The automobile industry had a great effect on Michigan in the early part of the 1900s. The way people traveled and lived changed because of a Michigan maker of automobiles, or cars. His name was Henry Ford. Ford did not build the first automobile, but he made many improvements in how automobiles were made.

The first automobiles were expensive because each was made by hand. Henry Ford believed that anyone—not just rich people—should be able to afford a car. Ford made cars cheaper to build by using a process called mass production. In **mass production**, many products that are alike can be made quickly and cheaply by using machines.

The assembly line made the mass production of automobiles possible. An **assembly line** is a line of workers and equipment along which a product moves as it is put together one step at a time. Henry Ford described the system by saying, "The man who puts in a bolt does not put on the nut; the man who puts on the nut does not tighten it." In 1908, without the assembly line, Ford's Model T car cost about $950 and took more than 12 hours to build. By 1927, after Ford had started his assembly line system, a Model T cost about $290 and could be made in less than two hours.

Ford wanted working people, like his own employees, to be able to afford to

(continued)

buy his cars. In 1914 the Ford Motor Company announced that it would pay assembly-line workers $5 a day. This was twice the pay of other factory workers at the time. Ford's ideas worked. In the 19 years that Ford made Model Ts, he sold more than 15 million of them.

Michigan quickly became the center of automobile manufacturing. In all, 270 automobile companies were started in Michigan between 1900 and 1909. General Motors was established in 1908 and manufactured cars in Flint, Detroit, Lansing, and Pontiac. By the 1920s General Motors was the biggest automobile manufacturer in the country. In 1914 Michigan produced almost four-fifths of the nation's cars and trucks. Today the Detroit area still produces more cars and trucks than any other place in the United States.

1. Which do you think was done first—installing the steering wheel or installing the gas tank? _____

2. Could the body of a car be installed before the engine? Explain. _____
_____
_____

3. At what point on the assembly line do you think the cars were painted?
_____
_____

4. Why did building cars on an assembly line save time and money?
_____
_____

5. What do you think are the advantages and disadvantages of doing only one job on an assembly line? _____
_____
_____
_____

ACTIVITY **23B**

SKILL: CATEGORIZE

Effects of the Automobile

DIRECTIONS: When you categorize information, you classify, or arrange, the information into similar groups so that it is easier to understand and compare. Read the paragraphs below. Then fill in the chart to categorize the ways the automobile changed life in the United States.

The automobile quickly became an important part of everyday life in Michigan and the rest of the United States. The automobile changed everything—from where people lived and worked to where people traveled.

With so many cars on the road, many new roads had to be built across the country. In 1900 the nation had only about 128,000 miles (about 206,000 km) of hard-surface roads. In 1924 there were more than 500,000 miles (805,000 km) of paved roads in the nation. Also as a result of the popularity of automobiles, the federal government built the interstate highway system in the 1950s.

With automobiles came new kinds of businesses. Across Michigan and the rest of the country, business owners opened gas stations, repair shops, and car dealerships.

Automobiles also changed the things Americans did for **recreation**, or for fun. Taking the car to eat at a drive-in restaurant or to see a drive-in movie became popular. Families began taking car trips on vacation. Motels, which are hotels for motorists, became common along highways. Often families drove to the country's national parks. This led the government to improve and expand the national park system.

Cars and more roads allowed people who worked in cities to live in the suburbs outside the cities. A **suburb** is a town or a small community built near a large city. People in the suburbs **commuted**, or drove to and from work in the cities. Suburbs grew, and **urban,** or city, areas across the country began to spread out.

While automobiles brought many positive changes to Michigan and the United States, the benefits have come at a price. Almost 3 million Americans have died in automobile accidents in the past 100 years. Also, the exhaust from millions of gas-burning engines has led to air pollution in urban areas, such as Detroit. Traffic jams, too, are common in most American cities.

THE AUTOMOBILE BRINGS CHANGES				
Transportation	Culture	Recreation	Economy	Environment

LESSON 24   THEME: INTERACTIONS

The Cereal Bowl of America

MAIN IDEA: In the 1890s Michiganians John Kellogg and W. K. Kellogg invented breakfast cereals, making Michigan an important food-processing center.

Another new industry in Michigan that began in the late 1800s was food processing. Because of an invention by two brothers, factories in Battle Creek began making breakfast cereals. The city soon became known as the Cereal Bowl of America.

In the 1890s John Kellogg, a doctor at a clinic in Battle Creek, and his brother, W. K., did experiments to make more healthful food for patients. The Kelloggs developed new kinds of grain, vegetable, and nut products. By accident in 1894 they invented wheat flakes. This was the first dry breakfast cereal.

In the late 1890s a man named C. W. Post stayed at the clinic. Afterward, he also experimented with health food. He created his own cereal called Grape Nuts and other food products. Post's products were very popular. Soon other people opened cereal companies in Battle Creek.

After his success with wheat flakes, W. K. Kellogg developed corn flakes. In 1906 he founded the Battle Creek Toasted Corn Flakes Company. The name was later shortened to the Kellogg Company.

An early advertisement

Early on, Kellogg spent a lot of money on advertisements. An **advertisement** is a public announcement that tells people about a product or an opportunity. As a result of advertising, Kellogg's Corn Flakes became the country's best-selling breakfast cereal.

Today the Kellogg Company is still the world's largest manufacturer of breakfast cereals. The company still has its headquarters in Battle Creek. Other food-processing factories, such as the Post Division of Kraft Foods and Ralston Foods, are located there as well.

RECALL
1. Who was W. K. Kellogg?
2. Why did Kellogg's Corn Flakes become the best-selling breakfast cereal?

CRITICAL THINKING—DISCUSS OR WRITE
3. **EVALUATION** Why did W. K. Kellogg start his company in Battle Creek? What happened to the city as a result?
4. **APPLY** How did Kellogg and Post change the way Americans eat breakfast?
5. **SYNTHESIZE** How does advertising help businesses sell their products?

ACTIVITY 24

SKILL: CHART AND GRAPH

From Corn to Corn Flakes

In 1894 Dr. John Kellogg and his brother, W. K., left some cooked wheat sitting out for almost two days at their clinic in Battle Creek. They decided to run the dried wheat through some rollers, which flattened the wheat kernels into small flakes. The Kellogg brothers then baked the flakes. These early wheat flakes were tough and tasteless. However, the Kellogg brothers soon improved the flakes and changed the way Americans ate breakfast.

Like the Kelloggs, cereal producers today and all other businesses need three kinds of resources to succeed. They need **human resources**, or workers and the skills they bring to the job. They need natural resources, such as water, minerals, and fuel. They also need **capital resources**, or the money, buildings, machines, and tools required to run a business. Together, these resources are called the **factors of production**.

DIRECTIONS: The diagram below shows how corn is made into cereal today. Study the diagram, and answer the questions that follow.

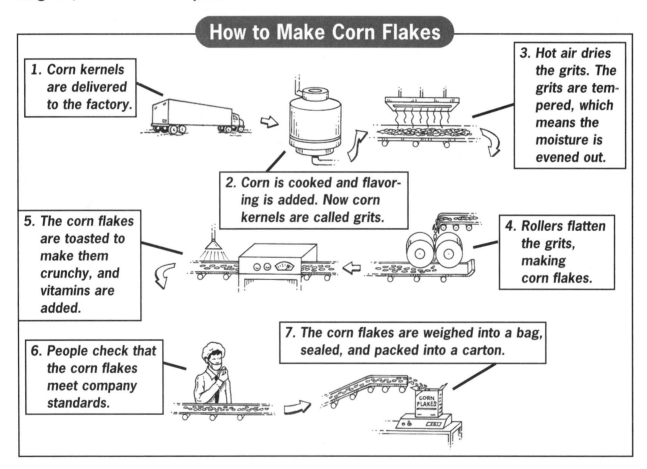

How to Make Corn Flakes

1. Corn kernels are delivered to the factory.
2. Corn is cooked and flavoring is added. Now corn kernels are called grits.
3. Hot air dries the grits. The grits are tempered, which means the moisture is evened out.
4. Rollers flatten the grits, making corn flakes.
5. The corn flakes are toasted to make them crunchy, and vitamins are added.
6. People check that the corn flakes meet company standards.
7. The corn flakes are weighed into a bag, sealed, and packed into a carton.

1. What happens before the corn is cooked? _____

_____

(continued)

2. What are grits? _____

_____

3. Why do you think the grits must be flattened before they are toasted?

_____

_____

4. Are vitamins added before or after the grits are dried? _____

_____

5. How is the cereal prepared for shipping? _____

_____

6. What do you think happens after the cereal is packed into cartons?

_____

_____

7. How might early cereal makers such as Kellogg and Post have applied some of

Henry Ford's ideas about mass production? _____

_____

8. Below is a list of some of the resources needed to make and deliver cereal to consumers. Categorize them as natural resources, human resources, or capital resources.

corn	rice	ovens	rollers
farmers	sugar	mills	inspectors
drivers	driers	cardboard boxes	trucks

NATURAL RESOURCES	HUMAN RESOURCES	CAPITAL RESOURCES

UNIT 5   MICHIGAN STATE ACTIVITY BOOK

UNIT 5

Practice Test

PART 1 SELECTED RESPONSE

DIRECTIONS: Beginning in the mid-1800s, Michigan experienced great industrial growth. Study the following table, and use it together with what you already know to answer the questions that follow.

Industry	Beginnings	Location	Products	New Technology
Farming	1830s	across Michigan	corn, wheat, and other crops	steel plow, reaper
Lumbering	1840s	Lower Peninsula, Saginaw Valley	lumber	narrow railroad tracks
Mining	1840s	Upper Peninsula	copper, iron ore, gold, coal, bromine, salt	dynamite, power drill
Manufacturing	late 1800s	Detroit, Grand Rapids, Battle Creek, Saginaw, Bay City	automobiles, furniture, cereal, stoves, railroad cars, chemicals, pharmaceuticals	assembly line

1. Which new technology played an important role in the farming industry?
   A. the assembly line
   B. narrow railroad tracks
   C. the reaper
   D. the power drill

2. Which industry played the GREATEST role in the growth of Michigan's cities?
   A. the mining industry
   B. the manufacturing industry
   C. the farming industry
   D. the logging industry

3. What was the MOST important effect of narrow railroad tracks on Michigan?
   A. It contributed to deforestation.
   B. It helped manufacturers deliver new cars.
   C. It brought immigrants to Michigan's farms.
   D. It slowed the development of industry.

4. Which of the following people helped develop an important new technology for the automobile industry?
   A. Cyrus McCormick
   B. John Deere
   C. C. W. Post
   D. Henry Ford

96 MICHIGAN STATE ACTIVITY BOOK

PART 2 EXTENDED RESPONSE PREPARATION

Interpret Graphs

DIRECTIONS: Study the line graphs below. One shows how the number of miles of railroad track in the United States increased from 1840 to 1900. The other shows how the number of factory workers in Michigan increased in the same period. Use this information and your social studies knowledge to answer the questions that follow.

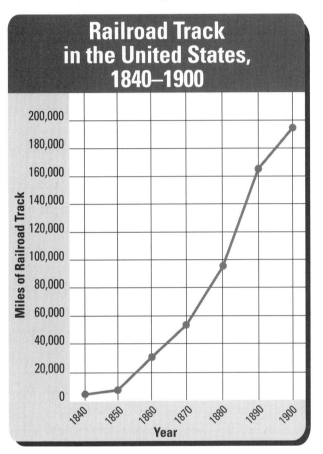

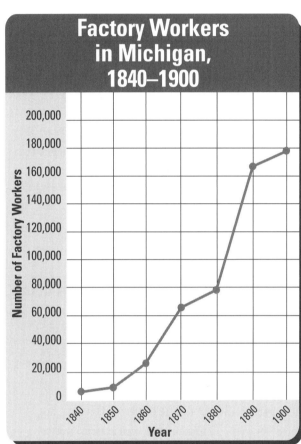

1. Do you think that the growth of railroads in the United States from 1840 to 1900 might have had an effect on the manufacturing industry in Michigan during the same time period?

   _____

2. Use information from the graphs, as well as your own social studies knowledge, to explain why you hold the opinion that you do.

   _____

   _____

   _____

   _____

LESSON 25

THEME: HUMAN EXPERIENCES

A New Century Brings Changes

MAIN IDEA: In the early years of the 1900s, Michigan was affected by war overseas and the movement of people to the North.

As Michigan's economy was growing in the early 1900s, events in other parts of the world would affect the state. In 1914 the Great War, later called World War I, broke out in Europe. Britain, France, and their allies went to war against Germany and its supporters. The United States eventually declared war against Germany on April 6, 1917.

Michigan factory workers made supplies that the military needed. Factories in Michigan and all across the United States were producing more goods than ever. However, the war took millions of American men away from their jobs in the factories. In addition, the number of immigrants dropped sharply because of the war. How could the United States produce enough goods when much of the workforce was gone?

One answer came from African Americans. Those who were not in the military moved by the tens of thousands from the South to the North to take jobs in the northern factories. So many African Americans moved north in the first half of the twentieth century, that the time is known as the Great Migration.

Between 1915 and 1930 as many as 500,000 African Americans left southern states. Some of these people headed to Michigan. During the years 1940 to 1943 alone, as many as 60,000 African Americans moved to Detroit.

Although African Americans found better lives in the North, they still faced discrimination (dis•krih•muh•NAY•shuhn). **Discrimination** is the unfair treatment of a group of people. Sometimes African Americans were not hired for jobs or allowed to rent apartments.

Southern white people also headed to Detroit and other cities in Michigan for new jobs. Disagreements between African American and white people sometimes took place as they competed for jobs and for places to live. In 1942 a disagreement about housing in Detroit turned into a riot. After the riot both African American and white people spoke out about ways to prevent further trouble.

RECALL
1. How did World War I affect Michigan?
2. In what direction did African Americans move during the Great Migration?

CRITICAL THINKING—DISCUSS OR WRITE
3. **CORE DEMOCRATIC VALUES: EQUALITY** Do you think African Americans who migrated to the North expected to be treated equally there? Explain your answer.
4. **EVALUATION** What do you think was a short-term effect of African American and white people from the South moving to Detroit? What was a long-term effect?
5. **INQUIRY** Do research to find out about the difficult times in the South that also led many African Americans to head to the North in the early 1900s. Then imagine you are living in the South during this time. Tell whether you would leave your family and friends in the South for the chance of a better life in Detroit.

ACTIVITY 25

SKILL: CHART AND GRAPH

Detroit and the Great Migration

DIRECTIONS: Use the information in the table to complete the line graph. Then use the line graph to answer the questions that follow.

Year	Number of African Americans in Detroit
1900	4,111
1910	5,741
1920	40,838
1930	149,119

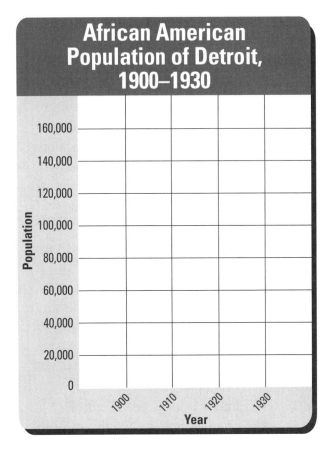

1. In which year was the African American population of Detroit

   the lowest? _____

   the highest? _____

2. By how much did Detroit's African American population increase from

   1920 to 1930? _____

3. During which ten-year period did the greatest change in Detroit's African American

   population take place? _____

4. How would you describe the line that you drew on the graph?

   _____

   _____

5. Based on the line graph, what general statement can you make about Detroit's African American population between the years 1900 and 1930?

   _____

   _____

LESSON 26

THEME: VALUES AND NEW IDEAS

The Great Depression

MAIN IDEA: From 1929 until the early 1940s, Michigan and the rest of the United States faced the worst economic crisis in the country's history.

From 1929 until the early 1940s, the United States suffered through the Great Depression. A **depression** is a time when there are few jobs and people have little money. The Great Depression began when the United States **stock market** "crashed." The stock market handles the buying and selling of stock, or shares in the ownership of a business. For most of the 1920s, stock prices kept going up. Many people borrowed from banks so they could buy more stock. In 1929 stock prices began to fall. On October 29, stock prices had fallen so low that the drop was called a crash. Almost everyone who owned stock lost money. Many people lost all their savings, and many banks closed.

By 1933 many people were out of work. Half of all people in Michigan who did not work on farms no longer had jobs. Without jobs, people in Michigan and across the country began spending less money. This meant that they bought fewer products. Many companies and factories had to close. The number of cars produced dropped from more than 5 million cars in 1929 to 1.3 million in 1932. The Ford Motor Company went from employing 128,000 workers in 1929 to just 37,000 workers in 1931.

Without jobs, many people did not have enough money to buy food or to pay rent. Charities opened soup kitchens to feed hungry people. Local governments helped, too. Mayor Frank Murphy of Detroit said, "No one in Detroit . . . must be allowed to go hungry, or cold or unhoused." The mayor asked people to start gardens so that they would have food. He also turned warehouses into places where people without homes could live.

In 1933 United States President Franklin D. Roosevelt started programs that created jobs for people who were **unemployed**, or out of work. Together, all the programs that helped people find work became known as the New Deal. About 100,000 young Michiganians worked for the Civilian Conservation Corps (CCC). Their duties included planting millions of trees and fighting forest fires. In addition, the Works Progress Administration (WPA) put 500,000 people in Michigan to work. The WPA workers built bridges, roads, and schools.

RECALL

1. When did the Great Depression take place?
2. What was the New Deal?

CRITICAL THINKING—DISCUSS OR WRITE

3. **EVALUATION** Michigan suffered more than most other states during the Great Depression. Why do you think this was so?
4. **ANALYZE** Why did the number of cars produced drop during the Great Depression? What happened to autoworkers as a result?
5. **CORE DEMOCRATIC VALUES: THE COMMON GOOD** How did Mayor Frank Murphy of Detroit try to help his community?

ACTIVITY 26

SKILL: ANALYZE POINTS OF VIEW

Living Through the Great Depression

DIRECTIONS: Read the quotations below by two people from Michigan who lived through the Great Depression. Then fill in the graphic organizer to analyze the speakers' points of view.

From an interview with Michigan resident Marie Beyne Gillis Tubbs:

"The business of my father (Theodore J. Beyne) was at a standstill. Since his hobby was playing the violin in the newly formed Grand Rapids Symphony Orchestra, he had time to search within himself for things to do. He began to compose beautiful music. . . .

"My first memory of hearing his music played was at the beginning of the Depression at the band shell at the city's John Ball Park. . . . How clearly I remember, out of the depths of dark feelings springing from closed banks and no work, the wonderful sensation that comes from something more than 'bread alone.' And I remember his pleased reaction (he was overwhelmed) at the audience's appreciation shown with lots of applause."

From an interview with Michigan resident Phyllis Bryant:

"In 1929 I was six years old, but I remember quite a few things from that era, especially growing up and never having too much. . . .

"Beans were a common meal and were often given to us by a farmer friend. What helped them along was the hot homemade bread. We usually had lots of homemade cookies and cakes, too. But it was kind of great, going to family reunions and eating . . . 'store bought' cookies and bread. . . .

"My dad was a carpenter and farmer and did lots of things to keep us going. We lived in the small village of Imlay City, close to a family that owned a cow. My dad milked her twice a day, fed her and cleaned the stall. In return we got two quarts of milk a day. With all the canning my mother did from our garden, our weekly grocery bill wasn't that big."

(continued)

DETAILS	POINT OF VIEW
Details from Marie Beyne Gillis Tubbs's interview	**Marie Beyne Gillis Tubbs's Point of View**
1. _____	_____
2. _____	
3. _____	
4. _____	

DETAILS	POINT OF VIEW
Details from Phyllis Bryant's interview	**Phyllis Bryant's Point of View**
1. _____	_____
2. _____	
3. _____	
4. _____	

LESSON 27

THEME: CONFLICT AND COOPERATION

The Rise of Unions

MAIN IDEA: During the Great Depression, workers formed labor unions as a way to gain better working conditions.

Those who found factory jobs during the Great Depression often faced tough working conditions. During these hard economic times, automobile factory owners wanted workers to work longer hours. Companies often cut workers' wages in order to stay in business. Also, workers had no control over how fast the factory's assembly line moved. The safety of the workers was sometimes given little thought.

Autoworkers thought they could bring about changes if they worked together. On August 26, 1935, they started a labor union called the United Automobile Workers (UAW). A **labor union** is a group of workers who take action to improve their working conditions.

The union workers wanted factory owners to talk with them about the changes that workers wanted. The owners refused. On December 30, 1936, the UAW ordered a sit-down **strike**, or a stopping of work, at a General Motors plant in Flint, Michigan. The strikers refused to work or to leave the plant. General Motors lost a great deal of money during the strike. In February 1937 General Motors managers agreed to talk with union representatives. The strike ended on March 12, 1937.

The UAW organized other strikes between 1937 and 1941. All the major automobile companies began speaking with the union. Soon, labor unions representing workers in other industries copied the UAW's methods. In time, so many people had joined unions that Michigan was nicknamed the "union state."

Autoworkers on strike

RECALL
1. What does *UAW* stand for?
2. When did the first major UAW strike against the General Motors plant in Flint begin and end?

CRITICAL THINKING—DISCUSS OR WRITE
3. **EVALUATION** Why did the UAW order a sit-down strike? What happened as a result?
4. **CORE DEMOCRATIC VALUES: THE PURSUIT OF HAPPINESS** Whose rights are affected during a strike?
5. **REFLECT** Why do you think factory owners would not talk with the union representatives at first?

ACTIVITY **27** SKILL: MAKE A THOUGHTFUL DECISION

Work for Change

DIRECTIONS: Read the situation described below. Then fill in the chart to help you make a thoughtful decision. Consider all the possible consequences, or things that might happen because of your action.

Situation: Imagine that you have a summer job, such as baby-sitting or washing cars. Like the factory workers who joined unions, you are happy to be earning money, but there are things about your job that you do not like. What will you do? Will you quit? ask for changes? live with things as they are?

1. What are three possible decisions you might make about your job?

A.

B.

C.

2. What are the positive and negative consequences of each decision?

A. POSITIVE:

NEGATIVE:

B. POSITIVE:

NEGATIVE:

C. POSITIVE:

NEGATIVE:

3. What is your final decision?

104 MICHIGAN STATE ACTIVITY BOOK UNIT 6

LESSON 28

THEME: COMPARISONS

World Events Affect Michigan

MAIN IDEA: Michigan soldiers overseas and factory workers at home helped the United States fight World War II.

In 1939, as Americans were struggling through the Great Depression, war broke out in Europe and Asia. At first, the United States stayed out of the war. Then in 1941 Japan bombed Pearl Harbor, an American naval base in Hawaii. Immediately, the United States joined the war. During the war the **Allies**, which included the United States, Britain, and France, fought the **Axis Powers** of Germany, Japan, and Italy.

All over the United States, people prepared for war. President Franklin D. Roosevelt ordered factories to make tanks, trucks, airplanes called bombers, guns, and bullets. Automobile factories in Michigan were quickly turned into factories that made materials to fight the war. Detroit made more wartime products than any other United States city. The Great Depression came to an end as factories across the country began producing wartime goods.

More than 670,000 people from Michigan served in World War II. Most were men, but some women served, too.

Because so many men marched off to war, new workers were needed in the factories. Women stepped in to do many of these jobs.

During the war, people at home often had to do without certain things. Shoes, sugar, and other goods were **rationed**, or limited. Michigan children helped the war effort by growing food in gardens called Victory Gardens. They also collected paper and tin to be **recycled**, or used again.

In 1945 World War II ended in victory for the Allies. Michigan's factories returned to making automobiles and other peacetime products.

During World War II, B-24 bombers were produced in a factory near Ypsilanti.

RECALL
1. How did World War II help bring the United States out of the Great Depression?
2. What kind of jobs did women hold during the war?

CRITICAL THINKING—DISCUSS OR WRITE
3. **EVALUATION** How do you think most people in 1942 felt about women working in factories? How do you think their feelings had changed by 1946?
4. **CORE DEMOCRATIC VALUES: PATRIOTISM** How did Michigan children show patriotism during World War II?
5. **INQUIRY** What else do you want to know about Michigan's role during World War II? Work with a partner to conduct research. Then share with the class what you learned.

ACTIVITY **28A**

SKILL: CHART AND GRAPH

The Arsenal of Democracy

DIRECTIONS: A picture graph uses small pictures or symbols to stand for the numbers of things. Read the passage below, and study the picture graph. Then answer the questions that follow.

Between 1939 and 1945, automobile companies supplied $50 billion worth of military goods to the United States government. Since Michigan was the center of the automobile industry, it made most of these military goods. In fact, during the war Michigan became known as the "Arsenal of Democracy." An **arsenal** is a place where military equipment is made or kept. By 1944, factories in the Detroit metropolitan area had received more than $12 billion worth of orders to make war supplies.

Key

$ = $2 billion

$ = $1.5 billion

¢ = $1 billion

▲ = $.5 billion

Automobile Industry Production of Military Items, 1939–1945

Kind of Item	Amount Spent
Airplanes and airplane parts	$ $ $ $ $ $ $ $ $
Trucks and other vehicles	$ $ $ $ $ $ $ ¢
Tanks	$ $ $ ▲
Weapons	$ $ $ $ ¢

1. For which kind of military item was the most money spent?

   _____

2. For which kind of military item was the least money spent?

   _____

3. About how many billions of dollars were spent on tanks?

   _____

4. About how much more money was spent on trucks and other vehicles than on weapons? _____

5. By 1944, did the Detroit metropolitan area receive about one-third or about one-fourth of the $50 billion worth of orders for war supplies?

   _____

ACTIVITY **28B**

SKILL: IDENTIFY FACT AND OPINION

Rosie the Riveter

DIRECTIONS: Read the passage below. Then complete the chart.

During World War II, many factories needed new workers. Many of the men who had worked in factories had gone off to fight the war. The United States government came up with ways to encourage women to work in factories. One way was to print posters that showed proud, strong women working in factories.

By 1944 about one-third of the factory workers in the United States were women. Soon women factory workers gained the nickname "Rosie the Riveter." A riveter is a person who attaches nuts and bolts that hold together planes, ships, and other objects covered with metal.

In Ypsilanti (ip•suh•LAN•tee), Michigan, a real-life Rosie the Riveter named Rose Will Monroe appeared in a government film. Women like Rose Will Monroe not only worked in factories but also took care of their homes and children. Still, they made less than men who did the same jobs. In 1941 women earned about $31.21 a week, while men earned $54.65 a week.

Although many women lost their factory jobs after the war, some found other jobs. After the war Rose Will Monroe drove a taxicab and owned a beauty salon.

FACT	OPINION BASED ON FACT
1. _____	1. Women were doing the same jobs as men and should have earned the same amount as men.
2. The government made posters showing women doing factory jobs.	2. _____
3. _____	3. Mothers who worked in factories had hard lives.

LESSON 29

THEME: INTERACTIONS

The Mackinac Bridge

MAIN IDEA: With the building of the Mackinac Bridge across the Straits of Mackinac, the Lower and Upper Peninsulas of Michigan were linked by road.

For years the Straits of Mackinac made it hard to travel between Michigan's Upper Peninsula and Lower Peninsula. As late as the 1950s, people had to cross the Straits on ferries. Finally, the leaders of Michigan decided to build a bridge across the Straits of Mackinac.

The bridge would have to withstand the area's winds and the ice that would form on it during winter. A suspension bridge would be able to do this. A **suspension bridge** is a bridge on which the roadway is hung, or suspended, from cables, or wire ropes, held in place by towers.

Work on the Mackinac Bridge began in 1954 and was finished in 1957. Nearly 5 miles (about 8 km) long, it became the world's longest suspension bridge.

Everyone who crossed the new bridge had to pay a toll. This money helped pay for building the bridge. The bridge was finally paid for in 1986. Today the money from tolls helps pay for bridge repairs.

The Mackinac Bridge was not the only important transportation project begun in Michigan in the 1950s. In 1956 the United States government decided to build more interstate highways. An **interstate highway** is a highway that goes through more than one state. Some of the interstate highways that run through Michigan include Interstate 94, Interstate 96, and Interstate 75. Most of the money used to build these highways came from taxes on gasoline.

The Mackinac Bridge

RECALL
1. Before 1957, what problem did people traveling between Michigan's Upper Peninsula and Lower Peninsula face?
2. Why did the designers of the Mackinac Bridge decide to build a suspension bridge?

CRITICAL THINKING—DISCUSS OR WRITE
3. **EVALUATION** What do you think were some short-term effects of the Mackinac Bridge? What do you think were the long-term effects?
4. **INQUIRY** With a partner, write questions about why a suspension bridge was the best kind of bridge to build over the Straits of Mackinac. Research the answers to your questions, and present the information to the class in an oral report.
5. **REFLECT** Do interstate highways affect your life? How would your life be different without interstate highways?

ACTIVITY 29

SKILL: MAP AND GLOBE

Michigan's Highways

DIRECTIONS: Use the Michigan road map to help you answer the following questions.

1. Which interstate highway goes through the Upper Peninsula?

   _____

2. Which United States highway follows part of the Lower Peninsula's shoreline along Lake Huron?

   _____

3. Which interstate highway would you take to go from Detroit to Grand Rapids?

   _____

4. What United States highway connects Marquette to Escanaba?

   _____

5. Imagine that your family is planning to travel by car from Ann Arbor to Lansing to Sault Sainte Marie. Describe the best route to take.

   _____

   _____

6. How would you travel by road from the Lower Peninsula to the Upper Peninsula without crossing the Mackinac Bridge?

   _____

   _____

UNIT 6

MICHIGAN STATE ACTIVITY BOOK 109

LESSON 30

THEME: VALUES AND NEW IDEAS

The Civil Rights Movement in Michigan

MAIN IDEA: Supporters of the Civil Rights movement in Michigan worked for equal rights for all.

As they had done in the early part of the twentieth century, African Americans continued to migrate from the South to the North after World War II. Often their lives were better in the North. However, African Americans in the North still faced unfair treatment.

Some people in Michigan and elsewhere in the country started working for **Civil rights**. Are the rights of citizens to equal treatment under the law. A Southern minister named Dr. Martin Luther King, Jr., spoke out against the unfair treatment of African Americans and became a leader in the Civil Rights movement. In 1963 King came to Detroit to lead a peaceful civil rights march. About 125,000 people joined him. King said that one day he hoped to see African Americans and white people "walking together hand in hand, free at last."

Dr. Martin Luther King, Jr.

In 1963 Michigan approved a new state constitution that promised equal rights to all. New national laws called for the integration of schools, restaurants, and public transportation. **Integration** is the including of all races as equals in a group.

Despite the changes brought on by the Civil Rights movement, some problems remained. Many African Americans still did not have the same opportunities to buy homes or receive an education as white people did. Banks rarely gave loans to African Americans to start businesses. Few African Americans were hired as police officers, firefighters, or teachers.

Civil Rights leaders called for peaceful protests against the continued unfair treatment of African Americans. Riots broke out, however, in Detroit and other Michigan cities in 1967.

RECALL
1. What are civil rights?
2. Who was Dr. Martin Luther King, Jr.?

CRITICAL THINKING—DISCUSS OR WRITE
3. **EVALUATION:** What do you think were some immediate effects of the riots of 1967? What were some long-term effects?
4. **CORE DEMOCRATIC VALUES: EQUALITY** How did Michigan's new constitution support equality?
5. **ANALYZE** How has the Civil Rights movement affected life in Michigan today? Give two examples.

ACTIVITY 30

SKILL: WRITING

The Freedom March

DIRECTIONS: A persuasive speech is a speech in which a person gives an opinion about a topic and tries to persuade listeners to agree with that opinion. Read about a persuasive speech that Dr. Martin Luther King, Jr., gave to the people of Detroit. Then, on a separate sheet of paper, write your own persuasive speech about your dreams for a better world. First, use the pre-writing form below to get started.

On June 23, 1963, civil rights supporters gathered in Detroit for the Great March to Freedom. Nearly 125,000 people marched down Woodward Avenue to Jefferson Avenue. The march ended at Cobo Hall. There, in front of thousands of people, Dr. Martin Luther King, Jr., gave an early version of his famous "I Have a Dream" speech. Here is just one part of that speech:

"This social revolution taking place can be summarized in three little words. They are not big words. . . . They are the words *all*, *here*, and *now*. We want all of our rights, we want them here, and we want them now. . . .

"And so this afternoon, I have a dream. It is a dream deeply rooted in the American dream. . . ."

The speech that Dr. Martin Luther King, Jr., gave in Detroit on June 23, 1963, was recorded on an album.

MY OPINION: _____

_____

MY REASONS:

1. _____

_____

2. _____

_____

3. _____

_____

UNIT 6

Practice Test

PART 1 SELECTED RESPONSE

DIRECTIONS: Read the paragraphs below, and answer the questions that follow.

> In early 1941 the Ford Motor Company began building the Willow Run Bomber Plant near Ypsilanti. The new factory was one of the largest factories in the world. It included not only space for building B-24 bomber airplanes but also a large airfield for testing them.
>
> People came from all over the United States to work in the Willow Run plant. Many of these workers were African Americans from the South. They came north looking for good jobs and more freedom. After World War II began, many male factory workers left to fight overseas. In their places, thousands of women began working at the Willow Run plant. As many as 42,000 men and women worked at the Willow Run plant at one time.
>
> At first it took a long time to produce a single plane. However, by 1943 the plant produced one B-24 bomber each hour. By the time the plant closed in 1945, at the end of World War II, it had produced 8,685 B-24 bombers. The bombers made at the Willow Run plant helped the Allies win the war. It is no wonder that this plant helped Michigan gain the nickname the "Arsenal of Democracy."

1. Why did African Americans from the South come to Willow Run?
   A. to build the bomber factory
   B. to fly bombers
   C. to work in the bomber factory
   D. to earn just enough money to return to the South

2. When men went off to fight overseas, who replaced them on the assembly lines?
   A. plant owners
   B. soldiers
   C. robots
   D. women

3. How did production change at the plant from its opening to its closing?
   A. Workers started making the bombers more slowly.
   B. Workers started making the bombers faster.
   C. Workers began making a bomber every minute.
   D. No change took place.

4. Why is Michigan referred to as the "Arsenal of Democracy"?
   A. Michigan's young men served in the United States Army.
   B. Michigan's citizens bought many bonds to support the war.
   C. Michigan's businesses recycled natural resources.
   D. Michigan's manufacturers produced important war materials.

(continued)

PART 2 EXTENDED RESPONSE

Use Core Democratic Values

DIRECTIONS: Read the following quotations. They are from a speech that President Franklin D. Roosevelt gave to Congress at the start of World War II. Today this speech is known as the Four Freedoms speech.

"In the future days, which we seek to make secure, we look forward to a world founded upon four essential human freedoms.

"The first is freedom of speech and expression—everywhere in the world.

"The second is freedom of every person to worship God in [his or her] own way—everywhere in the world.

"The third is freedom from want . . . everywhere in the world.

"The fourth is freedom from fear. . . ."

1. Do you believe that these words are as important to people today as they were for people during the time of World War II?

   _____

2. Use core democratic values to support your answer.

   _____
   _____
   _____
   _____
   _____
   _____
   _____
   _____
   _____

Lesson 31

THEME: VALUES AND NEW IDEAS

Michigan at the End of the Twentieth Century

MAIN IDEA: The people of Michigan faced hard times during the second half of the twentieth century.

In the summer of 2001, Detroit celebrated its 300th birthday. All over the city, people gathered to enjoy music, to watch shows, and to see history exhibits. They learned about their city's past, its many cultures, and its role in the American automobile industry. People also looked back at Michigan's recent past and the economic problems the state had faced.

Much of Michigan faced hard times in the second half of the 1900s. In the Upper Peninsula the state's mining industry had provided jobs for many years. However, miners did not do well in the second half of the 1900s. By the 1950s copper, iron, and other minerals in the Upper Peninsula had become harder to find. Mines closed, and workers lost their jobs.

Fortunately, the Upper Peninsula had another valuable resource—its natural beauty. After the Mackinac Bridge was completed in 1957, it was easier for people to travel to the Upper Peninsula to enjoy that beauty. Tourism soon became an important industry in the Upper Peninsula, creating new jobs for people there. Today, tourism continues to be important in the Upper Peninsula. People from cities all over the Middle West visit the Upper Peninsula to boat, fish, or enjoy quiet times. The tourists spend money on hotels, food, and tours.

The Lower Peninsula of Michigan also faced hard times in the last part of the twentieth century. In 1973 Michigan and the rest of the United States entered a **recession,** or business decline. The recession began when a group of countries stopped selling oil to the United States. This caused a drop in gasoline supplies in the United States. As a result, gasoline became more expensive. People who could afford new cars often bought cars made outside the United States. Most of those cars used less gasoline.

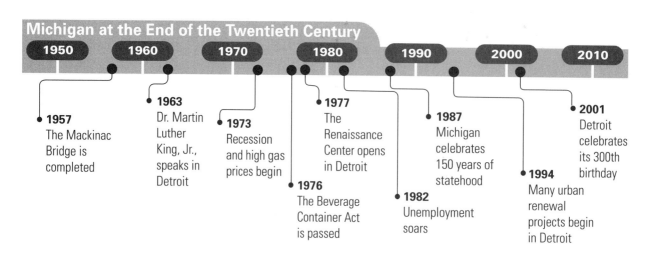

Michigan at the End of the Twentieth Century

- **1957** The Mackinac Bridge is completed
- **1963** Dr. Martin Luther King, Jr., speaks in Detroit
- **1973** Recession and high gas prices begin
- **1976** The Beverage Container Act is passed
- **1977** The Renaissance Center opens in Detroit
- **1982** Unemployment soars
- **1987** Michigan celebrates 150 years of statehood
- **1994** Many urban renewal projects begin in Detroit
- **2001** Detroit celebrates its 300th birthday

(continued)

The recession hit Michigan cities that had close ties to the automobile industry, such as Flint, especially hard. With fewer people buying American cars, thousands of autoworkers in Michigan lost their jobs. By March 1982 Michigan's unemployment rate was nearly twice that of many other states. Because fewer people in Michigan were working, fewer people paid taxes to the state. As a result, Michigan had less money for services such as police protection and road repairs. Many cities in the state seemed to be crumbling, and many people there began to lose hope. Residents and businesses moved out of the cities to the suburbs.

Then in the 1990s the United States economy grew stronger. Michigan cities began to make a comeback. From 1994 to 2002, people and businesses spent more than $12 billion in Detroit. New building projects such as Comerica Park—home of baseball's Detroit Tigers—and Ford Field—home of football's Detroit Lions—brought jobs. These projects also brought a new pride to the city. By the end of the 1990s, the number of people in Detroit who were without jobs was lower than it had been in several decades.

Michigan's leaders today hope that the state will not again face the kinds of economic problems that it struggled to solve in the 1970s and 1980s. They have worked to create a diverse economy in the state. A **diverse economy** is an economy that is based on many kinds of industries. Michigan's leaders are working to bring to the state more businesses that are not tied directly to the automobile industry. They believe that a more diverse economy will lead to a successful future for Michigan.

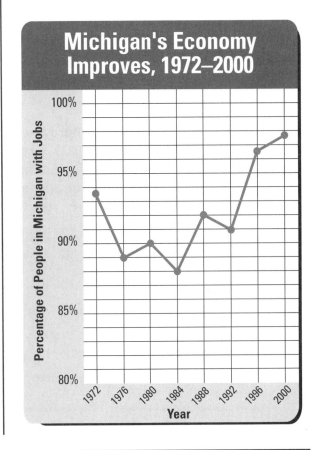

RECALL
1. What resource helped people living in the Upper Peninsula find new kinds of jobs in the second half of the 1900s?
2. What caused problems for Michigan's automobile industry in the 1970s?

CRITICAL THINKING—DISCUSS OR WRITE
3. **Synthesize** Why might having many different kinds of businesses make Michigan better able to handle hard economic times?
4. **Analyze** Why might businesspeople in the Upper Peninsula be interested in protecting the environment?
5. **Reflect** What do you think would have been hardest about living in Michigan during the recession?

ACTIVITY 31A   **SKILL: SOLVE A PROBLEM**

A City Grows Again

In the 1970s and 1980s, it seemed that there was little hope for Detroit. Many people there were out of work. Buildings and roads in the city were falling apart. During these decades, more than 300,000 people moved out of the city. Community leaders had to do something.

In the 1970s a group of businesspeople planned a building project in downtown Detroit that they hoped would **revitalize** (ree•VY•tuh•lyz), or bring life back to, the city. The planners called the project the Renaissance (reh•nuh•SAHNTS) Center. *Renaissance* means "revival." They believed that the center, which would have offices, theaters, restaurants, and stores, would attract many people to downtown Detroit.

In 1974 Coleman Young became Detroit's first African American mayor. He served as mayor until 1994. One of his main goals was to continue to improve Detroit. During Young's time as mayor, an arena and other large structures were built in Detroit. Over time, downtown Detroit improved.

DIRECTIONS: Think of a plan for improving part of your own community, just as the leaders of Detroit did in the 1970s and 1980s. First, copy the graphic organizer below on a separate sheet of paper, and fill it in. Then, on another sheet of paper, write a business letter that you might send to the mayor or another leader in your community, explaining your plan. A business letter has five parts: heading, greeting, body, closing, and signature. It is more formal than a personal letter.

ACTIVITY **31B**

SKILL: SEQUENCE

The Michigan Beverage Container Act

In the 1970s landfills in Michigan were filling up with empty beverage containers. A **landfill** is an area where garbage is placed and covered with earth. Other discarded containers littered roads. Manufacturers had to use more resources for new containers.

In 1976 Michigan's leaders passed the Beverage Container Act. This law required people who bought certain kinds of beverages to pay a ten-cent deposit on each beverage container. The law also stated that when people returned an empty container, they would get their deposit back. Returned containers would then be recycled.

By encouraging recycling, the Beverage Container Act reduces litter in Michigan and saves the taxpayers money. It also keeps many containers from being dumped in and filling up landfills.

DIRECTIONS: The Beverage Container Act encourages recycling. On the flow chart, write the recycling steps listed below in the correct order. Then answer the question.

People return their containers and collect their deposits.
Empty containers are recycled, or used again.
People buy beverages and pay a deposit on some containers.
Stores gather and send away empty containers.

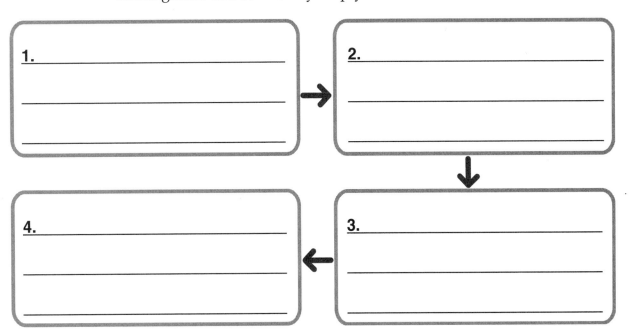

Why do you think Michigan's leaders decided to create the Beverage Container Act?

_____

_____

UNIT 7

MICHIGAN STATE ACTIVITY BOOK 117

LESSON 32

THEME: HUMAN EXPERIENCES

Who Lives in Michigan?

MAIN IDEA: The population of Michigan is made up of people of many different heritages.

About how many people lived in Michigan in 2000? If you guess about 10,000,000, you are right. The United States census for the year 2000 showed that 9,938,444 people called Michigan home in that year. A **census** is a count of all the people who live in a place.

The results of the 2000 census also showed that Michigan has a diverse population. Many different groups of people call Michigan home.

The diverse people of Michigan have a variety of heritages. A person's **heritage** includes his or her way of life, set of customs, and beliefs that come from the past and continue today. Each heritage adds some of its culture to Michigan. The culture of a people includes their language, art, music, clothing, crafts, food, holidays, and religion. Most people in Michigan have a European heritage. These Michiganians trace their roots back to such countries as Germany, the Netherlands, Italy, Greece, Poland, and France.

Heritages from many other parts of the world can also be found in the state. About one out of every seven people in Michigan has African ancestors. An **ancestor** is a family member from long ago. Some Michiganians have ancestors from many parts of Asia, including China, South Korea, Japan, Vietnam, India, and Pakistan.

People of Native American heritage also live in the state. Most Native Americans in Michigan make their homes in cities. However, some live on reservations, including a Potawatomi reservation in Hannahville and an Objibwa reservation near Sault Sainte Marie.

In recent years many Hispanics, or people from Spanish-speaking countries, have made Michigan their home. In fact, Hispanic heritage is the fastest-growing heritage in the state. Today about 325,000 people of Hispanic heritage live in the state of Michigan.

Some Heritages of Michigan's People	
Heritage	Population
Arab	116,300
Chinese	33,190
Czech	62,000
Dutch	480,800
English	988,600
French	488,800
French Canadian	191,900
German	2,028,200
Hungarian	98,000
Indian	54,600
Irish	1,068,900
Italian	451,000
Mexican	220,800
Norwegian	85,800
Polish	854,800
Russian	71,000
Scotch-Irish	130,200
Scottish	224,800
Subsaharan African	51,400
Swedish	161,300

Census 2000 figures

(continued)

More and more Arab Americans also have come to live in the state. Arab Americans are people who trace their heritage to countries in Southwest Asia where Arabic languages are spoken. Today more Arab Americans live in Michigan than anywhere else in the United States.

No matter what their heritage, most people in Michigan speak English. Many also speak another language. More than 50 different languages can be heard in Michigan on any one day. These include Spanish, German, Chinese, Polish, Arabic, Greek, and Korean.

Michigan is also a place of many religions. Most people in Michigan are Christians. Other religions in Michigan include Judaism, Islam, Buddhism, and Hinduism. Michigan's religious traditions add even more to the state's rich mix of cultures.

Michigan's Population	
Race*	Population
White	7,966,053
African American	1,412,742
Native American	58,479
Asian	176,510
Hispanic or Latino	323,877
Other Race	132,514
Ages	Population
Under 5 years	672,005
5 to 19 years	2,212,060
20 to 54 years	4,972,322
55 to 74 years	1,505,919
75 years and over	576,138
Total Population	**9,938,444**

Census 2000 figures
* Total number under Race is higher than Total Population. Some people responded as belonging to more than one race.

RECALL

1. About how many people live in Michigan?
2. What is the fastest-growing heritage in Michigan?

CRITICAL THINKING—DISCUSS OR WRITE

3. **SYNTHESIZE** The names of many Michigan cities are based on words from languages other than English. Why do you think this is so?
4. **REFLECT** Why do you think people of the same heritage often live together in the same community? Why do you think someone might move to a city where few people of his or her heritage live?
5. **CORE DEMOCRATIC VALUES: DIVERSITY** How does the diversity of Michigan's people provide benefits for the state?

Activity 32A

SKILL: CHART AND GRAPH

Newcomers to Michigan

Michigan's population is becoming more diverse every day. Each day immigrants from other countries arrive in Michigan to live. Today, more than half a million people who live in Michigan were born outside the United States.

Immigrants to Michigan today bring their cultures with them, just as immigrants have in the past. Their languages, customs and traditions, foods, music, and ideas become a part of Michigan's rich cultural mix. Immigrants to Michigan come from many different places. However, they share common goals—to make life better for themselves, their families, and their communities.

People in Michigan Born in Other Countries	
Place	Number of People
Europe	156,988
Asia	209,416
Africa	16,735
Pacific Islands	2,000
Latin America	88,704
Northern North America	49,659

DIRECTIONS: Make a bar graph showing immigration to Michigan. Use the information in the table above to create your graph.

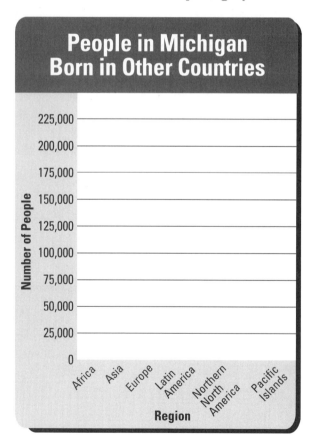

1. How many more people in Michigan were born in Asia than in Europe?

   _____

2. Which group of immigrants in Michigan was nearly twice as large as the number of immigrants from northern North America?

   _____

3. In 1990 Michigan's foreign-born population was 355,000. By about how much has the state's immigrant population increased? What conclusion can you make about Michigan's immigrant population in the future?

   _____

   _____

ACTIVITY **32B** SKILL: INQUIRY

Community Heritage

The city of Holland, Michigan, was started in the mid-1840s by Dutch immigrants. The Dutch are people from the Netherlands, a country in Europe. Sometimes people call the country Holland. That is how the Michigan city got its name. Early Dutch settlers named it after their home country. For many years most of the residents were people of Dutch background. They celebrated many Dutch customs and traditions and held special events, including an annual tulip festival.

Slowly the population of the city of Holland began to change. Today about one-fifth of its people are Hispanic. A number of other people trace their heritage to Asia or Africa. The customs of these newer immigrants have taken their place beside the Dutch customs and celebration. Of course, the city of Holland still holds its yearly Tulip Time Festival, honoring its Dutch heritage.

Examples of Dutch culture, such as windmills, can be seen in Holland, Michigan.

DIRECTIONS: Interview a family member to find out about your heritage. Use the library or the Internet to learn about the heritage of other people in your community. Then complete the table below.

Where my ancestors came from: _____

Languages my ancestors spoke: _____

Customs and traditions of my family: _____

_____

_____

Origin of the name of my community: _____

_____

_____

Year my family came to the community we live in now: _____

Heritage celebrations held in my community: _____

_____

UNIT 7 MICHIGAN STATE ACTIVITY BOOK 121

LESSON 33

THEME: INTERACTIONS

Where Do People Live in Michigan?

MAIN IDEA: In the past most people in Michigan lived in rural areas. Today most people in Michigan live in urban areas.

Most people think of Michigan as a state of large cities, such as Grand Rapids, Flint, Ann Arbor, Detroit, and Lansing. There is a good reason for this. About two-thirds of the people of Michigan live in urban areas, such as cities and suburbs. In fact, more than 2 million Michigan citizens live in the state's ten largest cities.

Michigan has not always been an urban state. Until about 1920 most Michiganians lived in rural areas. There people worked on farms, in the mining or logging industries, or in small businesses.

As Michigan's manufacturing industries grew, people in Michigan moved from the countryside to the cities. Farmers from southern Michigan took jobs in the furniture factories of Grand Rapids. Loggers from the Upper Peninsula moved south to Detroit, Flint, and Dearborn to work in automobile factories. European immigrants and African Americans from the southern United States also moved to the cities to work in these factories.

By 1950 Detroit had reached its highest population ever—1,849,568 people. Then another migration began. People in cities began to dream of owning their own homes in places that were quieter and less crowded. They moved to suburban communities, or smaller towns outside city centers.

As Michigan's cities grew, more and more suburbs rose up around them. People began to speak of cities and their suburbs as metropolitan areas. **Metropolitan areas** are huge land areas of connected cities and suburbs. Today most people in Michigan live in metropolitan areas.

However, not everyone in Michigan lives in metropolitan areas. Most of Michigan's metropolitan areas are in the southern and eastern parts of the state. In other parts of Michigan, people live in smaller cities, in towns, and on farms.

The rural parts of Michigan have a much lower population density than the urban areas of the state. **Population density** is the average number of people living in a square mile of land. Wayne County, which includes part of Detroit, has a population density of about 3,400 people per square mile. In contrast, Oceana (oh•shee•A•nuh) County, along

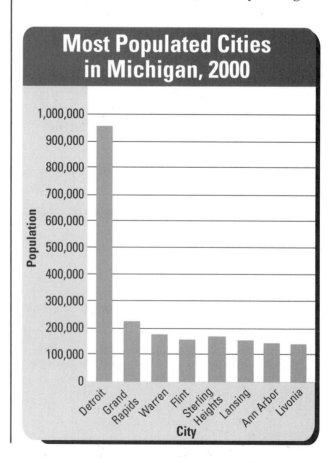

(continued)

122 MICHIGAN STATE ACTIVITY BOOK

UNIT 7

the shore of Lake Michigan, has a population density of only about 42 people per square mile.

Many of today's residents of Michigan's rural areas live there because of the natural resources and the jobs that these resources provide. In northern Michigan the mining and logging industries still offer good jobs. Other people there have found work in the growing tourism industry.

Michigan's land itself offers opportunities for Michigan's rural population. Farming is still a way of life in many rural communities. In the Upper Peninsula and the northern part of the Lower Peninsula, potatoes are an important crop. Along the shores of Lake Michigan lies the state's fruit belt. In this area many kinds of fruit grow well. Apples, cherries, and blueberries are the leading crops. Each year Michigan's farmers grow more than 1 million pounds of fruit.

Most residents of rural Michigan are pleased with the lives they lead. Clean air, a low crime rate, and little noise and traffic all help make communities in rural Michigan good places to live.

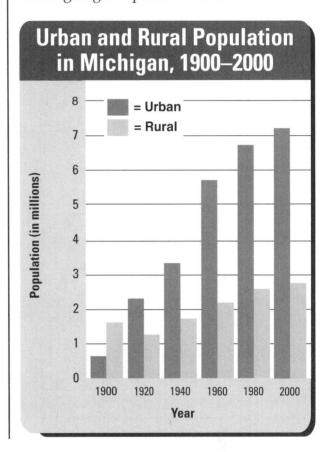

RECALL
1. About how much of Michigan's population lives in urban areas?
2. Why do some people like living in rural Michigan?
3. How does the population density of Wayne County compare to the population density of Oceana County?

CRITICAL THINKING—DISCUSS OR WRITE
4. **ANALYZE** Why might residents of Michigan's suburbs one day want to return to city centers to live?
5. **EVALUATION** Why do you think urban areas in Michigan have a more diverse population than rural areas do?
6. **SYNTHESIZE** What are some reasons that people live where they do in Michigan?

ACTIVITY 33

SKILL: MAP AND GLOBE

Michigan's Population

DIRECTIONS: Use the population density map below to help you answer the following questions.

1. In what part of Michigan is population density highest?

   _____

2. In what part of Michigan is population density lowest?

   _____

3. Near what place in the Upper Peninsula is population density highest?

   _____

4. What is the population density around Traverse City?

   _____

5. Which city has a higher population density—Kalamazoo or Bay City?

   _____

6. Where in Michigan do you think population density will increase most in the next ten years? Explain.

   _____

   _____

   _____

   _____

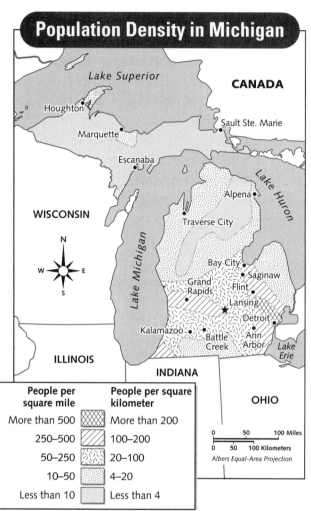

124 MICHIGAN STATE ACTIVITY BOOK

UNIT 7

LESSON 34

THEME: VALUES AND NEW IDEAS

Michigan's Economy Today

MAIN IDEA: People in Michigan produce a wide variety of goods and services.

Michigan's economy affects everyone in the state, young and old. As you have read, an economy is the way people use resources to meet their needs. Such needs include food, clothing, housing, transportation, education, and more. These needs are important to you as a student, just as they are to your parents and other relatives.

All over the world, people meet their needs in different ways. This means that countries around the world have different kinds of economies. These economies can be divided into three main kinds.

In a **traditional economy,** people use the natural resources around them to meet their needs. These resources might include animals and plants for food and clothing. They might also include wood for fuel and building. Long ago, Native American groups in Michigan, such as the Ojibwas and the Menominees, had traditional economies.

In a **command economy,** the central government controls all the resources. It also decides what products the people should make and how much the products should sell for. Cuba is one of the few

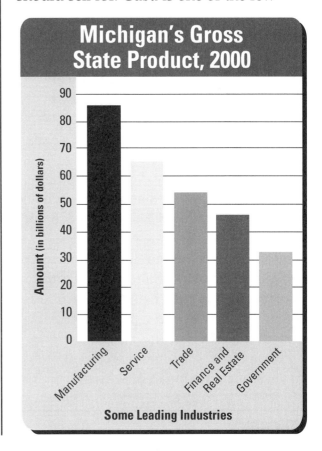

Employment in Michigan Industries	
Industry	Number of People Employed
Service	1,676,078
Manufacturing	1,045,651
Trade	702,574
Construction	278,079
Finance and Real Estate	246,663
Transportation and Communications	191,799
Government	167,731
Agriculture and Mining	49,496

(continued)

UNIT 7   MICHIGAN STATE ACTIVITY BOOK 125

countries in the world that still has a command economy.

In a **market economy,** the people—not the government—own and control businesses. People also decide what they will buy and where they will work. Michigan, along with the rest of the United States, has a market economy. This kind of economy is also called a free enterprise system.

In a market economy, people often have two different roles. As **producers** they make goods and offer services for sale to meet the needs of others. As **consumers** they buy goods and services to meet their own needs.

A market economy depends on the idea of supply and demand. The products and services offered by producers are the **supply.** The wants and needs of consumers are the **demand.** If consumers demand more of a certain product, producers will make more of it. If consumers demand less of a certain product, producers will make less of it.

Because of consumer demand, businesses in Michigan offer many different kinds of products and services. Michigan products include automobiles, furniture, foods, and houses. Services include child care, teaching, hair cutting, and home building.

The demand for products and services in Michigan provides jobs for the state's **labor force,** or workers. Look at the table titled "Employment in Michigan Industries" on page 125. Notice that jobs in Michigan can be divided into categories, such as service, trade, manufacturing, and agriculture and mining. The table shows that many people in Michigan work in service, trade, and manufacturing industries. Fewer people work in farming and mining.

Together, all these workers produce many different products and services. Each product and each service is worth something. In other words, each has a value. The total value of all the products and services produced in a state in a single year is the **gross state product (GSP).** In the year 2000 Michigan's GSP was about $325 million—the ninth-largest GSP of all the states.

RECALL
1. What kind of economy does Michigan, like the rest of the United States, have?
2. Which industry in Michigan employs the most people?

CRITICAL THINKING—DISCUSS OR WRITE
3. **REFLECT** Why do you think it is important that people in Michigan and the rest of the United States are able to choose the kind of work they do?
4. **SYNTHESIZE** Michigan's GSP is spread across many different industries. Why do you think industries in Michigan are so diverse?
5. **CORE DEMOCRATIC VALUES: LIBERTY** Which kind of economy offers more liberty to the consumer—a command economy or a market economy? Explain.

ACTIVITY **34A**

SKILL: CATEGORIZE

The Service Industry in Michigan

The service industry is the fastest-growing part of Michigan's economy. In the **service industry,** workers are paid to provide services, or do things, for other people, rather than to make products, such as automobiles. Doctors, teachers, and grocery store clerks are just a few examples of workers in the service industry. Today more people in Michigan work in the service industry than in any other industry. About four out of every ten people in the state have a service job. Each year the service industry contributes more than $60 billion to Michigan's economy—or about one-fifth of its GSP.

Health care is a service industry.

DIRECTIONS: Complete the table below by listing examples of the kinds of jobs that people have in each kind of service industry.

SERVICE INDUSTRY	EXAMPLES OF JOBS
Health Care	
Education	
Tourism	
Transportation	
Food Service	
Retail Trade	
Information	
Finance and Real Estate	

ACTIVITY **34B** SKILL: SEQUENCE

The Automobile Industry in Michigan Today

DIRECTIONS: Use the information in the paragraphs below to complete the flow chart.

Although service jobs make up a large part of Michigan's economy, automobile manufacturing continues to be important to the state. In the year 2001 the automobile industry employed more than 255,000 people and paid about $45 billion in wages. These workers produced more than 3 million vehicles in 13 automobile plants in Michigan.

What do you think happens in one of these state-of-the-art factories? First, automobile parts, such as steel, tires, and batteries, arrive at a factory's intake center. These parts, along with parts made right there in the factory, then go to the assembly room. Most automobile manufacturers today use two assembly lines to build cars, one for the body of the car and one for its chassis (CHA•see). Body parts include doors, windows, and seats. Chassis parts include engines, transmissions, and brakes. Workers on the body line and the chassis line put the cars together, using high-tech equipment. Automobile companies even use robots to do certain jobs. Near the end of the assembly process, the two lines come together, and workers finish the car.

After a car is put together, it is inspected to make sure it is of high quality. Then a few of the cars are selected for testing on a track by specially trained drivers. If the cars pass the tests, they are sent to showrooms to be sold.

HOW CARS ARE MADE

1. Employees receive or make parts in the company's intake center.

2. Workers take the parts to the correct assembly line—the body line or the chassis line.

3. _____

4. _____

5. _____

6. _____

7. The automobiles are shipped to showrooms.

LESSON 35 THEME: INTERACTIONS

Michigan Looks to the Future

MAIN IDEA: Michigan's future depends on a strong economy, good schools, and a clean environment.

Michigan today is very different from what it was years ago. The state has gone through great changes over the years, and it will continue to change in the future.

The people of Michigan want a bright future for their state. Working together to make this possible are businesspeople, teachers and principals, government leaders, and ordinary citizens.

A strong and diverse economy is a key to success for the state. Today individuals and groups in Michigan are working to help businesses grow and to attract new businesses to the state. One such group is the Michigan Economic Development Corporation.

You and the education that you receive are very important to Michigan's future.

Michigan's future also depends on education. The Michigan Department of Education has set high learning goals for its elementary, middle, and high school students. In school districts across the state, parents, teachers, principals, and students are working to make sure that these goals are met.

Important, too, for a bright future is Michigan's environment. The state depends on its environment in many ways. Businesses and individuals rely on natural resources to meet their needs. Tourists visit Michigan to see its natural beauty. State leaders have taken many steps to make sure that Michigan's environment stays clean. One of these steps is the formation of the Great Lakes Program to Ensure Environmental and Economic Prosperity. The goals of this program are to clean up pollution, restore wetlands, and keep the Great Lakes beautiful.

Over the past 300 years, Michigan has experienced good times and bad. The hard work of Michiganians today will help Michigan become an even better place tomorrow.

RECALL
1. What group is working to attract new businesses to Michigan?
2. What are three keys to success for Michigan?

CRITICAL THINKING—DISCUSS OR WRITE
3. **SYNTHESIZE** How will meeting Michigan's education goals make the state a better place in which to live in the future?
4. **EVALUATION** How will meeting learning goals help you right away? How will it help you in the future?
5. **INQUIRY** What questions might you ask if you wanted to learn more about protecting Michigan's environment now and in the future?

Activity **35A**   **Skill:** Predict a Likely Outcome

Predict Michigan's Future

DIRECTIONS: You have learned a lot about Michigan's past and present. Now use the chart below to make predictions about Michigan's future. A prediction is a statement of what you believe will happen in the future based on the way things are now. For each topic listed in the table, describe what it is like now. Then tell what you think it will be like ten years from now. Be sure to give evidence to support your prediction.

MICHIGAN'S FUTURE			
Topic	What It Is Like in Michigan Today	Prediction	Evidence to Support Prediction
Urban Life			
Rural Life			
Education			
Environment			

ACTIVITY **35B**

SKILL: USE PRIMARY AND SECONDARY SOURCES

Discover the History of Your Community

DIRECTIONS: Your community today and what it will be like in the future are affected partly by your community's past. Work in a group to prepare a display about the history of your community. Be sure to include at least six paragraphs of writing, four drawings or photographs, and two quotations. Use the guidelines below to help your group plan its display.

PROJECT GUIDELINES

1. Make sure that each team member has a part of the project to complete. Decide what your display will look like and what kinds of information will be included.

2. Conduct research. At the library, try to find both primary sources and secondary sources. Primary sources might include local newspapers from the past, public records, and journals of people who lived long ago. Secondary sources might include books or Web sites about your community and its history.

3. Gather and organize your information. Remember to focus on the most interesting and most important people and events in your community's history. Include information about how the community got its start, changes through the years, and the people and cultures that make up the community.

4. Work together to prepare your final display. Proofread all written words. Make sure that every photograph or drawing has a caption that identifies it. Check that each quotation is followed by the name of the person who said it and when he or she said it.

5. Prepare a list of the primary sources and secondary sources that your group used. Attach it to the back of the display.

6. Use the evidence about your community that you gathered to make and tell predictions about your community's future. Show your display in a presentation.

UNIT 7

Practice Test

PART 1 SELECTED RESPONSE

DIRECTIONS: Study the map of Michigan below. Then use it to help you answer the questions that follow.

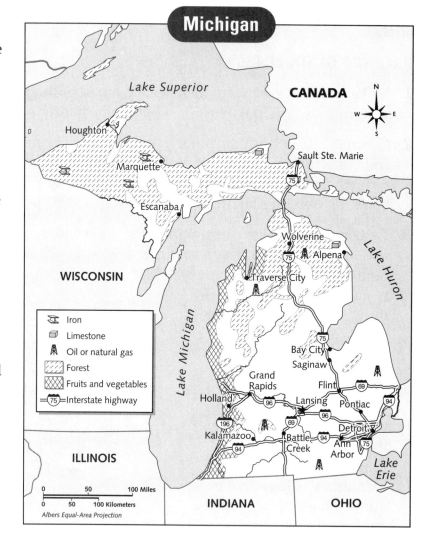

1. Which interstate highway plays the most important role in the economy of the Upper Peninsula?
   A. I-96
   B. I-94
   C. I-69
   D. I-75

2. Which interstate highway would fruit and vegetable farmers near Holland use to get their produce to Lansing?
   A. I-75
   B. I-69
   C. I-96
   D. I-94

3. Where in Michigan would you find the most cherry trees?
   A. the western part of the Upper Peninsula
   B. the western part of the Lower Peninsula
   C. the eastern part of the Upper Peninsula
   D. the eastern part of the Lower Peninsula

4. Near which community in Michigan would you find the most forestlands?
   A. Kalamazoo
   B. Wolverine
   C. Battle Creek
   D. Flint

(continued)

PART 2 EXTENDED RESPONSE PREPARATION

Use Core Democratic Values

DIRECTIONS: Read the following material about a public policy issue—whether to revise the Beverage Container Act. Use the information together with what you already know to complete the two tasks that follow.

> ### Revising the Beverage Container Act
>
> Michigan's Beverage Container Act requires consumers to pay a 10-cent deposit on certain beverage containers. They get the deposit back when they return the empty containers. Because of the law, people in Michigan litter less, recycle more, and save energy.
>
> The law costs Michigan's state and local governments and private businesses a great deal of money. Each year, people turn in some containers for which they did not pay a deposit. (They may have bought them in another state.) The state pays $10 million to cover the cost of these illegally returned containers. Beverage companies have to pay one cent per container to handle them. This costs the companies $50 million a year.
>
> Some people believe that the container law should be eliminated. They say that most people recycle cans and bottles with their trash and that most containers could be put in recycling programs for free. Many others say the container law should be kept. They believe that it has helped keep Michigan not only cleaner but safer as well.

Imagine that this question was asked of students in a Michigan school: "Do you think the Beverage Container Act should be eliminated?" The circle graphs below show possible answers.

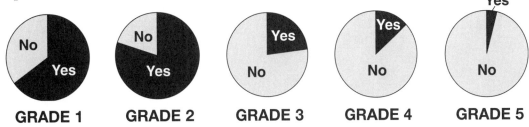

GRADE 1 GRADE 2 GRADE 3 GRADE 4 GRADE 5

Task I: Interpreting Information

Based on the graphs, which of the following statements is TRUE?
- **A.** Students in all five grades want to eliminate the Beverage Container Act.
- **B.** Students in all five grades want to keep the Beverage Container Act.
- **C.** Most students in Grades 1 and 2 want to eliminate the Beverage Container Act.
- **D.** Most students in Grades 4 and 5 want to eliminate the Beverage Container Act.

Task II: Taking a Stand

1. Do you think Michigan should eliminate the Beverage Container Act? On a separate sheet of paper, explain why you hold that opinion. Use core democratic values to support your answer.

Questions About History

Ask yourself these questions as you read.

1. WHAT happened?

2. WHEN did it happen?

3. WHO took part in it?

4. HOW and WHY did it happen?

History

People	Events	Ideas

Main Idea and Supporting Details

| Supporting Detail | Supporting Detail |

Main Idea

| Supporting Detail | Supporting Detail |

Compare and Contrast

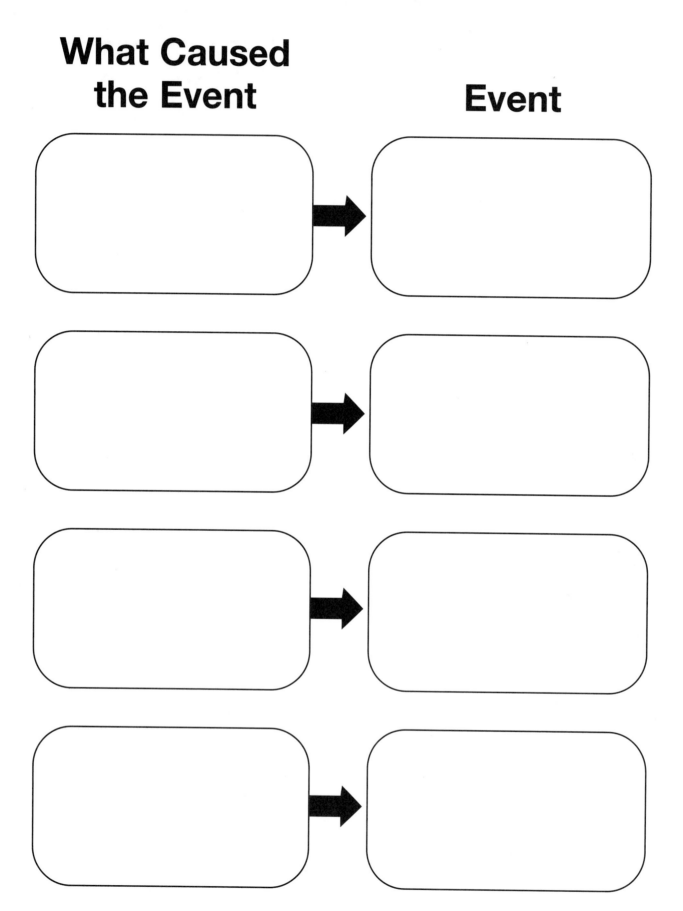

Sequence

Event | **Order**

Event	Order

THINKING ORGANIZERS

Categorize

Time Line

Michigan

The United States

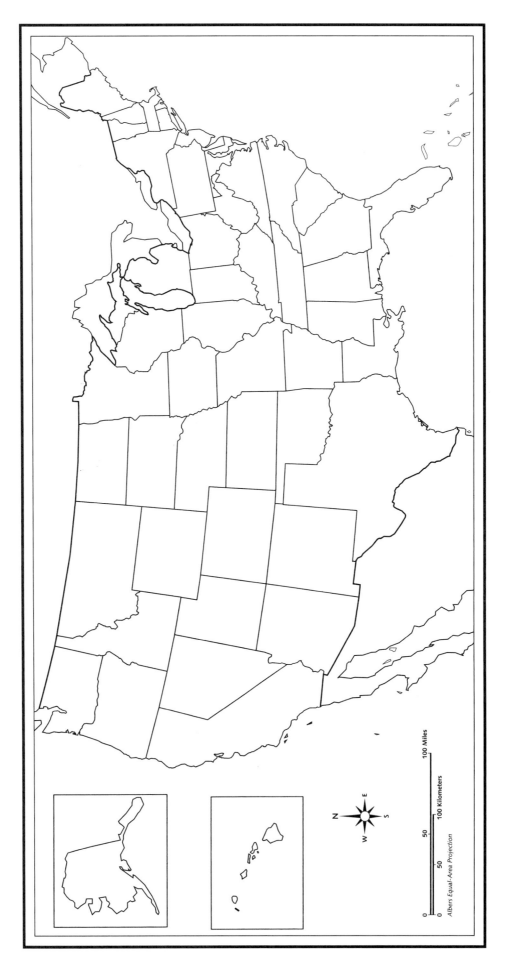

THINKING ORGANIZERS

MICHIGAN STATE ACTIVITY BOOK 143